The National Computing Centre develops techniques, provides services, offers aids and supplies information to encourage the more effective use of Information Technology. The Centre co-operates with members and other organisations, including government bodies, to develop the use of computers and communications facilities. It provides advice, training and consultancy; evaluates software methods and tools; promotes standards and codes of practice; and publishes books.

Any interested company, organisation or individual can benefit from the work of the Centre – by exploring its products and services; or in particular by subscribing as a member. Throughout the country, members can participate in working parties, study groups and discussions; and can influence NCC policy.

For more information, contact the Centre at Oxford Road, Manchester M1 7ED (061-228 6333), or at one of the regional offices: London (01-353 4875), Bristol (0272-277 077), Birmingham (021-236 6283), Glasgow (041-204 1101) or Belfast (0232-665 997).

Do You Want to Write?

Could you write a book on an aspect of Information Technology? Have you already prepared a typescript? Why not send us your ideas, your 'embryo' text or your completed work? We are a prestigious publishing house with an international reputation. We have the funds and the expertise to support your writing ambitions in the most effective way.

Contact: Geoff Simons, Publications Division, The National Computing Centre Ltd, Oxford Road, Manchester M1 7ED.

Computer-related Mathematics and Statistics

Brian Patterson

PUBLISHED BY NCC PUBLICATIONS

British Library Cataloguing in Publication Data

Patterson, Brian
    Computer-related mathematics and statistics.
    1.   Mathematics – For computer sciences
    I.   Title
    510

ISBN 0-85012-665-7

First published in 1988 by:

NCC Publications, The National Computing Centre Limited, Oxford Road, Manchester M1 7ED, England.

Typeset in 10pt Times Roman by H&H Graphics, Blackburn; and printed by Hobbs the Printers of Southampton.

ISBN 0-85012-665-7

Contents

Introduction

This book is specifically designed to meet the needs of those who are studying on the NCC Threshold or Diploma courses, but is equally suited to the needs of any students on introductory courses, who are required to have an understanding of the important ways in which mathematics and data processing interrelate. There exists a deep-rooted tradition that implies that it is necessary to be 'good' at mathematics in order to be 'good' at data processing. Whatever the basis for this tradition it is patently untrue, although there is evidence that many of the characteristics that are exhibited by skilled mathematicians, notably in respect of logical and analytical ability, are also needed for programmers and for systems analysts.

In this text, importance is given to what *needs* to be known and how to apply techniques, rather than to the mathematical theory *behind* those techniques. In an ideal world, the theory and the practical should go together but for the world in which we live it is probably more important to establish the practicalities for the benefit of all rather than both theory and practice for the minority who can appreciate both.

Equally important in this text is the need to produce copious examples, both worked in the body of the chapters and also left for the reader to attempt. These examples are, in most cases, set in a data processing context since it is going to be in such a context that the reader needs to apply them. To undertake problems which relate to file sizes, to computer word-length and to down-time analysis, provides the reader with the best possible opportunity to see the relevance of what they are learning. Too many texts already exist which fail to relate the subject-matter to the disciplines to which they are to be applied and, because they *are* general-purpose, they frequently fail to provide what the reader requires.

Whilst it is possible that this text can be studied on its own it is far better that it should be used in conjunction with lectures based on the material used, since this should ensure that even more examples are demonstrated. Additionally the lecturer is able to provide a more versatile treatment of the material, varying the presentation to suit the audience, a flexibility it is hard to produce in any text.

1 Arithmetic Calculations

1.1 CALCULATIONS INVOLVING SIGNED INTEGERS

Anybody who was taught arithmetic at school cannot possibly have failed to have undertaken numerous calculations involving integers (or whole numbers) so there is little or nothing to be learned in respect of new ideas. However, it might be appropriate to be reminded of how to deal with the signs (+ and −) of the quantities involved:

(i)   If two quantities are multiplied together and their signs are *the same,* then the product will always be + , but if their signs are *different* then the product will always be − . (Note that +6 and 6 are totally interchangeable in this context):

$$\text{eg} \quad 6 \times +3 = +18$$
$$-5 \times -7 = +35$$
$$\text{but} \quad -3 \times +8 = -24$$
$$\text{and} \quad +7 \times -6 = -42$$

(ii)   If one quantity is divided by another then the rules are exactly the same as for multiplication:

$$\text{eg} \quad +36 \div 4 = +9$$
$$-45 \div -3 = +15$$
$$+30 \div -5 = -6$$
$$-28 \div +7 = -4$$

(iii)   If we try to add a negative number such as –5 the result is the same as if we had tried to subtract 5, thus:

$$\text{eg} \qquad 12 + \ -3 \ = 12 \ - 3 = 9$$
$$23 + \ -7 \ = 23 \ - 7 = 16$$
$$18 + (\ -5) = 18 \ - 5 = 13$$

Note that the (occasional) use of brackets is to ensure that the – and the 5 are kept together and does not alter the meaning.

(iv)   If we try to subtract a negative number such as –4 the result is the same as if we had tried to add 4, thus:

$$\text{eg} \qquad 17 - \ -3 \ = 17 + 3 \ = 20$$
$$26 - \ -4 \ = 26 + 4 \ = 30$$
$$19 - \ -12 = 19 + 12 = 31$$
$$63 - (\ -32) = 63 + 32 = 95$$

1.2   EXERCISES

Calculate each of the following, noting that if brackets are used to enclose an expression which can be itself calculated, then this expression must be first calculated before the rest is attempted.

(a)	8×15	(l)	$23 + -11$
(b)	$-7 \times \ -3$	(m)	$18 + -9$
(c)	$-8 \times +13$	(n)	$27 + (-6)$
(d)	$+6 \times -17$	(o)	$59 + (-32)$
(e)	$18 \times (\ -11)$	(p)	$28 - +8$
(f)	$48 \div 4$	(q)	$19 - +13$
(g)	$-55 \div -11$	(r)	$22 - +17$
(h)	$-27 \div +3$	(s)	$34 - (+11)$
(i)	$56 \div -7$	(t)	$14 - (+9)$
(j)	$+15 \div (-3)$	(u)	$3 \times -4 \times -2$
(k)	$14 + -4$	(v)	$-6 \times -3 \times -7$

(w) 29 + (–3 × –2)   (y)   18 – +3 + –7

(x)   +13 – (–3 × 2)   (z)   25 + –6 + –8

1.3 EXERCISES ON INTEGRAL CALCULATIONS IN A COMPUTER ENVIRONMENT

The following set of questions is designed to test simple applications of integer calculations, but based upon the situations which may exist in some part of a data processing environment.

1. A full box of line printer stationery contains 2000 sheets. How many sheets are contained in 37 boxes?

2. 203 cylinders on a disk pack exist for data storage. Of these, four have been used for control and system information and there are already three files stored which each occupy 23, 39 and 16 cylinders respectively. How many cylinders are still available for other purposes?

3. A processor operates at 8 mips (million instructions per second); how many seconds are needed for it to obey 144 million instructions?

4. A print buffer has a capacity of 132 characters. Four fields occupy respectively columns 12 to 29, columns 37 to 53, columns 68 to 80 and columns 93 to 108. How many columns remain unfilled?

5. A programmer produces, on average, 156 lines of tested code per day. How many lines does he produce in six weeks, working five days a week, except for two days when he was away on a course?

6. A file holds 12,500 records. Each record consists of seven fields which contain respectively 23, 16, 29, 14, 6, 12 and 9 characters.

    (a) How many characters are contained in the whole file?

    (b) If the file is to be held on a storage medium on which 2 Mbytes are available for its storage (1 Mbyte = 1,000,000 bytes and 1 byte holds one character of information) how many bytes are left over? (NB: Ignore inter-block gaps, header labels, etc.)

(c)   If the file header and trailer occupy a total of 178 bytes
and if there are 251 inter-block gaps each of which occupies
900 bytes, how much space is available on the storage
medium for other data?

1.4 CALCULATIONS INVOLVING FRACTIONS

There are a number of rules to remember when dealing with
fractional values and these may be expressed as follows:

(i)   When adding or subtracting fractions both need to be expressed
with the same denominator (the bottom part of a fraction)
before simplification can be undertaken. Hence:

$$2/3 + 5/6 = 4/6 + 5/6 = 9/6 = 3/2 \text{ or } 1^1/2$$

In this case the $2/3$ could not be added to the $5/6$ until each of
them had been expressed with the same denominator; this
would have to be calculated as the smallest number into
which both denominators, 3 and 6, could divide which would
therefore be 6. In order to convert the denominator 3 into 6 it
would have to be doubled. The same has to be done to the
numerator (the top part of a fraction) hence the 2 was
doubled to get 4. Having reached this stage the denominators
become the same so the two numerators, 4 and 5, can be
added together to get the 9. Subsequently both the numerator,
9, and the denominator, 6, can be divided by 3 to simplify the
result to $3/2$ which is the same as $1^1/2$.

$$eg \qquad 3/8 - 5/16 = 6/16 - 5/16 = 1/16$$

$$7/10 - 1/3 = 21/30 - 10/30 = 11/30$$

$$2/7 + 1/4 = 8/28 + 7/28 = 15/28$$

$$1/2 + 1/3 = 3/6 + 2/6 = 5/6$$

$$4/9 + 1/2 = 8/18 + 9/18 = 17/18$$

If one or more of the fractions is a mixed number such as $2^1/2$,
$3^1/8$, $7^2/9$, etc then it is possible to either treat the integer part
and the fractional parts separately so that:

$$eg \quad 2^1/2 + 3^1/3 = (2 + 3) + (1/2 + 1/3) = 5^5/6$$

$$3^1/8 - 1^1/15 = (3 - 1) + (1/8 - 1/15) = 2^7/120$$

or they can be converted into improper fractions so that $2^1/_2 = {}^5/_2$, $1^1/_{15} = {}^{16}/_{15}$, etc, and then proceed normally:

eg $2^3/_4 - 1^1/_2 = {}^{11}/_4 - {}^3/_2 = {}^{11}/_4 - {}^6/_4 = {}^5/_4$ or $1^1/_4$

$3^4/_7 + 2^1/_3 = {}^{25}/_7 + {}^7/_3 = {}^{75}/_{21} + {}^{49}/_{21} = $
$\quad {}^{124}/_{21}$ or $5^{19}/_{21}$

(ii) When multiplying two fractions together cancel wherever possible and then multiply the numerators together, likewise the denominators, to obtain the solution, hence:

eg $\quad {}^7/_8 \times {}^2/_{21} = {}^1/_8 \times {}^2/_3$ (on cancelling 7)
$\quad = {}^1/_4 \times {}^1/_3$ (on cancelling 2) $= {}^1/_{12}$

$\quad {}^{15}/_{24} \times {}^{36}/_{55} = {}^3/_{24} \times {}^{36}/_{11}$ (on cancelling 5)
$\quad = {}^1/_8 \times {}^{36}/_{11}$ (on cancelling 3)
$\quad = {}^1/_2 \times {}^9/_{11}$ (on cancelling 4) $= {}^9/_{22}$

If one or both fractions are mixed numbers they *must* be expressed as improper fractions first:

eg $\quad 1^1/_2 \times 2^2/_3 = {}^3/_2 \times {}^8/_3$
$\quad = {}^1/_2 \times {}^8/_1$ (on cancelling 3)
$\quad = {}^1/_1 \times {}^4/_1$ (on cancelling 2) $= 4$

$\quad 4^2/_7 \times 5^1/_3 = {}^{30}/_7 \times {}^{16}/_3$
$\quad = {}^{10}/_7 \times {}^{16}/_1$ (on cancelling 3) $= {}^{160}/_7$ or $22^6/_7$

Do note that with mixed numbers it is *not possible* to calculate their product correctly *by multiplying the integer parts and the fractional parts separately.* $4^2/_7 \times 5^1/_3$ is *not* $20^2/_{21}$!

(iii) When dividing one fraction by another, leave the first fraction alone, change the division sign to a multiplication sign and invert the second fraction, then proceed as for multiplication:

eg $\quad {}^8/_9 \div {}^2/_3 = {}^8/_9 \times {}^3/_2$
$\quad = {}^4/_9 \times {}^3/_1$ (on cancelling 2)
$\quad = {}^4/_3 \times {}^1/_1$ (on cancelling 3) $= {}^4/_3$ or $1^1/_3$

$\quad 2^3/_4 \div 1^1/_3 = {}^{11}/_4 \div {}^4/_3 = {}^{11}/_4 \times {}^3/_4 = {}^{33}/_{16}$
$\quad$ or $2^1/_{16}$

1.5   EXERCISES INVOLVING FRACTIONS

1. Calculate each of the following, simplifying your answer whenever possible:

   (a) $3/10 + 1/2$ (g) $4/9 \times 3/16$

   (b) $2/7 + 1\frac{1}{4}$ (h) $12/25 \times 15/16$

   (c) $2\frac{1}{8} + 3\frac{1}{9}$ (i) $1\frac{1}{4} \times 1\frac{1}{15}$

   (d) $7/8 - 1/2$ (j) $3/4 \div 1/2$

   (e) $4\frac{1}{3} - 1\frac{1}{6}$ (k) $7/10 \div 14/5$

   (f) $3\frac{1}{7} - 2\frac{3}{4}$ (l) $2\frac{3}{4} \div 2\frac{1}{16}$

2. In the magnetic tape library there are 480 tapes in all; of these four-fifths are 9-track tapes and the rest are 7-track tapes. How many 7-track tapes are there?

3. One half of the 72 staff in the computer department are male and one third of the rest are married. How many unmarried women are there in the department?

4. Over a 24-hour period the main processor was down for one third of the time; routine test jobs occupied a quarter of the time that was left and one third of the balance was then taken up with the systems testing of a new stock control system. How many hours were left for other work?

5. On an exchangeable disk pack there are 11 disks with 400 tracks on each surface. There are 13,500 characters per track and the transfer rate is 806,000 cps (characters per second):

   (i)   How many characters can be stored on the pack (bearing in mind that the two outermost surfaces cannot be used to record data)? Give your answer in Mbytes.

   (ii) If one third of the pack holds data on company product lines, how many Mbytes are left for other purposes?

   (iii) How long will it take to transfer 4,191,200 characters?

6. Three women in the data preparation section have keyboard speeds of 8500, 9200 and 9800 characters per hour respectively.

If they work together on a job which contains 123,750 characters how long will it take them to complete it? How many characters will each woman have input?

1.6   CALCULATIONS INVOLVING DECIMALS

When working with decimals the approach is very similar indeed to that adopted when working with integers except for the positioning of the decimal point. The position can be easily determined if, when adding or subtracting, all values are so lined up underneath one another that the decimal points are aligned, thus:

eg   31.76 + 2.754 + 197.2 should be set out as:

$$\begin{array}{r} 31.76 \\ 2.754 \\ 197.2 \\ \hline 231.714 \\ \hline \end{array}$$

51.82 − 24.975 should be set out as:

$$\begin{array}{r} 51.82 \\ 24.975 \\ \hline 26.845 \\ \hline \end{array}$$

In multiplication it is necessary to count up how many figures there are after the decimal point in each of the numbers being multiplied and add these together; this result is directly the number of figures after the decimal point in the product. Hence we have to multiply the numbers together paying no attention whatsoever to the positions of the decimal points, putting the decimal point into the result afterwards on the basis of the earlier calculation.

eg   Multiply 3.75 by 16.207. There are $2 + 3 = 5$ figures after the decimal point in the answer so multiply 375 by 16207 to get 6077625; then put in the decimal point five positions from the right-hand end to get the answer as 60.77625.

In cases of division, suppose a quantity A needed to be divided by a second quantity B. Move the decimal point in B to the right until the number has been made an integer, counting up how many places it had to be moved; now move the decimal point in A by the same number of places to the right. Divide the new A by the new B in the usual way. Hence:

> eg   Divide 37.6 by 0.47. Move both decimal points two places to the right so that now we are dividing 3760 by 47, giving an answer of 80. This answer is in fact the same whether we are dividing 37.6 by 0.47 or 3760 by 47.

1.7   EXERCISES INVOLVING DECIMALS

1.   Add together 27.65, 1964.3 and 11.217.

2.   Subtract 59.73 from 107.3.

3.   Multiply 21.34 by 2.1.

4.   Divide 3.87 by 0.3.

5.   A consultant charges $53.50 an hour. What will he charge for 3.5 hours work?

6.   Three files require, respectively, 2.3 Mbytes, 1.74 Mbytes and 3.165 Mbytes of storage space.

   (i)   What is their total requirement?

   (ii)   How much space is left on a 10 Mbyte disk after storing these files?

7.   Two numeric fields have to be multiplied together in a program. Field A has four digits before and two after the decimal point, whereas field B has three digits before and three after the decimal point. How many digits will there be in the product AB:

   (i)   Before the decimal point?

   (ii)   After the decimal point?

1.8   RATIOS

If we were to divide up the staff in a computer department into operational and development it might be found that there are seven on the operational side and eight on the development side. We could then speak of dividing the 15 staff up in the ratio 7:8. If, however, it turned out that there were 14 operational and 16 development staff then the ratio remains exactly the same although this time we might speak of dividing up a staff of 30 in the ratio 7:8. The ratio provides a comparison between the respective sizes of the operational and development teams and, as such, like a fraction it ignores absolute values.

Two fractions could easily be established on the basis of this ratio; since $7 + 8 = 15$, we might view the '7' as holding seven shares out of a total of 15 available or a fraction of $7/15$ which represents the operational staff expressed as a fraction of the total staff. Again, the fraction would be the same whether considering seven people out of a total of 15 or 14 out of a total of 30. The fraction $8/15$ can also be extracted in a similar manner to express the number of people in the development team as a fraction of the total staffing of the department. Ratios may be viewed therefore as being closely akin to fractions.

On re-examination of the department it may be found that of the 15 staff, five are female and 10 are male. This means that the ratio of females to males is 5:10 or 1:2 (since both sides can be 'cancelled down'); it also means that the ratio of males to females is 2:1. If 12 of the 15 staff were non-graduates and three were graduates, then the ratio of graduates to non-graduates will be 3:12 or 1:4.

If the task of generating 500 lines of code was shared between two programmers Farooq and Adrian in the ratio 2:3, then Farooq is expected to handle two 'shares' out of the five (2+3) available so that he is expected to contribute $2/5$ of the work or $2/5 \times 500 = 200$ lines, against the $3/5 \times 500 = 300$ for Adrian. Note that, as a check, $200:300 = 2:3$.

Sometimes, increasing the size of memory may be referred to in the ratio 2:1; this should be taken as meaning that there is a comparison being made between the 'new' size, represented by two 'shares' and the 'old' size, represented by one 'share'. Hence the 'new' is twice as large as the 'old'. If the old were 8 Mbytes, then the new will be 16 Mbytes. Likewise if increasing in the ratio 5:4 with an 'old' memory size of 8 Mbytes, the new would have to be 10 Mbytes, since each 'share' of which the old has four, will be worth $8/4$ or 2

Mbytes so the new will have to be 5 × 2 Mbytes or 10 Mbytes.

Occasionally a statement such as 'The ratios of profits between the three divisions P, Q and R of a company were as 3:2:7 and totalled £24,000' may be encountered. Clearly this is a way of comparing three quantities, the profits of P, Q and R in a single statement; if we were to add up the 3, the 2 and the 7 a total of 12 'shares' is obtained which leads at once to the conclusion that each share must be worth £24,000/12 or £2,000. Hence, since division P gets 3 shares out of the 12 it will secure a profit of 3 × £2,000 or £6,000; likewise Q gets 2 shares or 2 × £2,000 = £4,000 and R gets 7 × £2,000 = £14,000. It is well worth verifying that the three figures obtained, £6,000, £4,000 and £14,000 do indeed total £24,000.

1.9   EXERCISES INVOLVING RATIOS

1.   Computer time is shared between Production and Maintenance in the ratio 5:1. How many hours do Maintenance get over the course of a 5-day (12 hours a day) week?

2.   Of the 120 people employed in a computer bureau, 84 are aged 21 or above. What is the ratio of those aged '21 or above' to those 'less than 21' in this bureau?

3.   The cost of listing paper is shared between the DP department and a user department in the ratio 4:3. How much does the user department pay when the total cost of listing paper is £476?

4.   If the size of the programming section was increased in the ratio 5:3 from its present size of 15 staff, how many staff will be employed in the future?

5.   In a large civil engineering company computer department the processor time is shared between commercial applications, engineering applications and system software activities in the ratios 8:6:1. How many hours are used for engineering applications out of a 90-hour week?

6.   If the amount of space in the computer department used for housing the machine room area is decreased in the ratio 5:4 because of the use of smaller processing equipment, what size will now be given over to the machine room if it had previously occupied 95 square metres?

2 Percentages and Indices

2.1 PERCENTAGES

A percentage is only a special kind of fraction. If we need to calculate 7% of a number then that is exactly the same thing as working out $^7/_{100}$ of that number. Hence 7% of £500 is $^7/_{100} \times$ £500 = £35, after cancellation. As a general rule therefore x% of anything is exactly the same as $^x/_{100}$ of that quantity.

One of the commonest encounters most people have with percentages is in respect of the interest rates offered by banks or building societies in which people may have put their savings. Hence if I have £150 in a savings account and it pays 8% per annum interest then I will get an extra $^8/_{100} \times$ £150 or £12 interest after one year so increasing the amount I have there to £150 + £12 or £162. When that £162 is left for a second year, with the same rate of interest I will get $^8/_{100} \times$ £162 or £12.96 interest this time, so that there is now £162 + £12.96 or £174.96 in my savings account. The process whereby the interest increases each year in this way is referred to as *compound interest* and this is indeed how most accounts work.

However, even with building society accounts the interest rate may be quoted as a more awkward figure such as $8^3/_4$% per annum. This can easily be dealt with by calling the rate 8.75% (using a decimal value instead of the fraction) in which case $8^3/_4$% of £260 is the result of working out $^{8.75}/_{100} \times$ £260 which comes to £22.75. Alternatively, $8^3/_4$% may be expressed as an improper fraction $^{35}/_4$ and this can then be worked out as $^{35}/_{400} \times$ £260 which gives exactly the same answer of £22.75.

23

We could be asked to express one value as a percentage of another. In this case, it is first expressed as a fraction, then multiplied by 100 and the percentage is immediately found. Hence, if faced with expressing 12 as a percentage of 300 we calculate the fraction, $^{12}/_{300}$, and multiply by 100 to give $^{12}/_{300} \times 100$ or 4%. In the same way, if asked to express 60 as a percentage of 800 it would be found as $^{60}/_{800} \times 100 = 7^{1}/_{2}\%$.

Quite frequently one hears of computer hardware being an asset which depreciates; in other words its value reduces with the passage of time. Hence we may find that some hardware is reckoned to depreciate at, say, 15% per annum. Thus, if the initial cost was £80,000 the amount of depreciation in the first year will be $^{15}/_{100} \times$ £80,000 = £12,000 and so the value of the equipment has reduced to £68,000.

If the documentation which should accompany any computer system is examined we are likely to find statements such as 'Growth rate . . %'. This is included to enable the designer to state by what percentage he or she expects the volume of data processed by the system to increase each year. The file size may be of 8,500 records initially and if expected to grow by 4% per annum, it will increase by $^{4}/_{100} \times 8,500$ or 340 records in the first year to a size of 8,840 records; in the second year it is expected to grow by $^{4}/_{100} \times 8,840 = 353.6$ or 354 records (to the nearest integer) hence making the file 9,194 records in size.

2.2   PERCENTAGE CHANGE

If a value were to change, say from 25 to 28, you might be asked to find the percentage change it had undergone. In this case first find what actual change has taken place, which is $28 - 25$ or 3, and then express this as a percentage *of the original figure*, 25; hence the percentage change would be $^{3}/_{25} \times 100$ or a 12% increase (since the value has gone *up*). In exactly the same way we can talk about a percentage profit in which case the actual profit is expressed as a percentage of what the article(s) cost in the first place (the cost price); hence if we buy something for £20 and sell it for £23.50 an actual profit of £3.50 has been made or a percentage profit of $^{3.50}/_{20} \times 100 = 17^{1}/_{2}\%$.

Problems on percentage profit get more troublesome when the cost and the selling prices are quoted in respect of different quantities of articles, as in the following example:

Books are bought in bulk at a cost of £6.50 for a dozen but are sold at £0.75 each; calculate the percentage profit made. (It does not matter whether it is calculated with the prices each or the prices for a dozen so long as it is consistent; here it is easier to quote the prices per dozen in each case.) Hence the selling price is £9 per dozen so that the profit made is £2.50 per dozen. The percentage profit is $2.50/6.50 \times 100 = 5/13 \times 100 = 500/13 = 38.5\%$ (correct to one place of decimals).

2.3 EXERCISES INVOLVING PERCENTAGES

1. Calculate 12% of £750.

2. A departmental budget is increased from £5,500 by 8%. What is its new value?

3. A microcomputer sells at £1,700 but is then reduced in price by 7%. What is its new price?

4. If the sum of £500 was invested in a savings account which paid 6% per annum interest and if the interest was added to the account at the end of each year with nothing drawn out from the account at any stage, then how much would be in the account after the end of:

    (i)   the first year;

    (ii) the second year;

    (iii) the third year?

5. If £2,000 was invested in a high yield account which paid $9\frac{1}{4}\%$ per annum interest, how much interest would be earned in the first year?

6. The price of a computer package is £350 and the price of a microcomputer is £1,800. Express the price of the package as a percentage of the price of the microcomputer.

7. The equipment in a large computer installation is valued at

£1,300,000 but it is depreciating at the rate of 12% per annum based upon its value at the start of the year. What is its value after:

  (i)   One year?

  (ii) Two years?

8. A file occupies 65 Kbytes of secondary storage but is assumed to grow at the rate of 5% per annum based upon its size at the start of the year. How much storage will it require after:

  (i)   One year?

  (ii) Two years?

9. A programmer who is paid £400 per month receives an increase and as a result now gets paid £430 per month. What percentage rise did he get?

10. The cost of a box of floppy disks goes down from £16 to £13. What is the percentage decrease in price?

11. I can buy floppy disks at £12 for a box of 10, but I sell them at £1.25 each. What is my percentage profit on the sales?

12. The usual price for some public domain software is £36 for any five titles but as a special promotion they are being sold at £39 for any seven. What is the percentage reduction in price?

2.4 INDICES

Although the use of indices (also called powers or exponents) will be dealt with more thoroughly in later chapters concerned with algebra, they can occur in straightforward arithmetic situations from time to time so they merit at least a brief introduction at this stage. Rather than have to write down 7 times 7 this may be abbreviated to 7^2 (read as 7 squared or 7 to the power of 2) in which case 2 is called an index (plural 'indices') and indicates how many of the 7s are to be multiplied together. For additional examples, consider:

Hence if we wish to write $5 \times 5 \times 5$ this could instead be written as 5^3 (read as 5 cubed or as 5 to the power of 3) but both have the same value, 125. The index in this case is 3, since there are three 5s being multiplied together. For additional examples, consider:

$$4 \times 4 \times 4 \times 4 \times 4 = 4^5 = 1024$$

$$7 \times 7 \times 7 \times 7 = 7^4 = 2401$$

$$2 \times 2 \times 2 \times 2 \times 2 \times 2 \times 2 = 2^7 = 128$$

It is possible however, for an index to be either zero, negative, or fractional even if these do not appear to meet the criterion of 'how many' of a certain value are being multiplied together. Without going into the formal proofs each of these can be defined as follows:

(i)   Any value which has a zero index is *always* equal to 1. Hence $8^0 = 1$, just as 3^0 or 5^0 or 12^0.

(ii)   For a negative index simply ignore the negative sign but place the whole expression as the denominator of a fraction in which the numerator is 1, thus:

$$8^{-1} = {}^1/_8$$

$$4^{-3} = {}^1/_{4}{}^3 = {}^1/_{64}$$

$$5^{-2} = {}^1/_{5}{}^2 = {}^1/_{25}$$

$$2^{-5} = {}^1/_{2}{}^5 = {}^1/_{32}$$

(iii)   Any fractional index represents a 'root', thus:

$$4^{1/2} = \text{square root of 4 (which is 2) since } 2^2 = 4$$

$$27^{1/3} = \text{cube root of 27 (which is 3) since } 3^3 = 27$$

$$32^{1/5} = \text{fifth root of 32 (which is 2) since } 2^5 = 32$$

2.5   EXERCISES INVOLVING INDICES

1.   Rewrite each of the following using indices and in each case give also the result of evaluating the expression:

   (a) $4 \times 4 \times 4 \times 4 \times 4$

   (b) $2 \times 2 \times 2 \times 2 \times 2 \times 2 \times 2 \times 2$

   (c) $5 \times 5 \times 5 \times 5$

   (d) $3 \times 3 \times 3 \times 3 \times 3 \times 3 \times 3$

2.   Find the value of each of the following:

   (a) 6^0

(b) 3^{-2}

(c) 4^{-1}

(d) $8^{1/3}$

(e) $25^{1/2}$

(f) $1024^{1/10}$

3 Number Base Representations and Elementary Calculations

3.1 INTRODUCTION

Everywhere in the world people are used to calculating and to representing numbers in what is called the decimal (or sometimes denary) number system. This system is based upon our own physical characteristic of having ten fingers and the fact that this has been the traditional basis of our counting systems. There have existed other number systems based upon numbers such as 60 (in ancient Mesopotamia) or 20 (in France where the word for 80 today is still 'quatre-vingts', literally four twenties); but these systems have only restricted use today.

Computers, however, do not have ten 'fingers' and so we have to make use of the characteristics of electronics and this means the use of 'two-state' devices, as in the case of an electrical signal being low (0, zero) or high (1, one), a gate being closed (0) or open (1), a punchcard or paper-tape containing a frame with no hole in it (0) or a hole in it (1).

The number system used in these two-state devices is called binary and there are two other systems closely related to it, called octal and hexadecimal. In the pages which follow all three will be investigated and the way they relate to one another and to the decimal system examined.

3.2 BASES IN COMMON USE AND SYMBOLS USED

Each number system has what is called a base, which indicates the number of different symbols used in that system. In the case of the decimal system, this base is 10 since there are ten different symbols used to show quantities in that system (these being the digits 0, 1, 2, etc, up to 9). Should we require to represent a quantity less than 10 this can be done by the use of only one digit, whereas larger quantities need the use of two or more digits to represent them. From what has already been said about the binary system, the base must be two, with only the digits 0 and 1 available, so to represent quantities of two or over requires the use of two or more digits, making it at first appear quite clumsy.

The octal number system has 8 as its base and uses the symbols 0, 1, 2 up to 7 only; in this case two or more digits are needed to represent quantities such as eight or above. The hexadecimal number system has 16 as its base and since the traditional decimal system has only ten digits available we have to introduce the letters A, B, C, D, E, F to stand for the 'digits' ten, eleven, twelve, thirteen, fourteen and fifteen. This does appear as rather strange at first but gives a logical way of providing the 16 'digits' needed for the system (Figure 3.1).

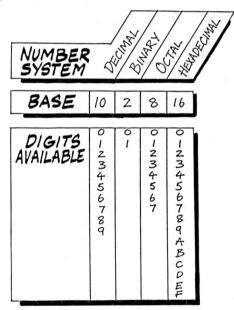

Figure 3.1   Number Base Representations

3.3   POSITIONAL VALUES

Although having so far referred to the bases and to the digits used within each number system, they have not been related to the value of any particular quantity.

3.3.1 Decimal

If, for example, we consider the number 258 in the decimal system, then the 8 represents the number of units, and since the digit 5 occupies a position one place to the left of the units column it represents the number of tens. With 10 as the base of the decimal system, each position further to the left represents a value ten times greater at each move; hence the 2 represents the number of 'ten times ten' or hundreds; thus 258 has a value given by 2 hundreds plus 5 tens plus 8 units.

It follows therefore, that the place values in the decimal system are (moving from the right to the left of a whole number) units, tens, hundreds, thousands, tens of thousands and so on. For this reason the 2 in the example does not have a value of one quarter of the 8, since it represents 200 not 2 units.

Place value depends upon the base used, so the second position from the right represents tens only because ten is the base; and the third position from the right represents hundreds only because 'the base times the base' is a hundred. So, if we consider the four quantities 5, 57, 513 and 5728, the 5 has a different value each time since in the first it represents 5 units, in the second 5 tens (50), in the third 5 hundreds (500) and in the fourth 5 thousands (5000).

3.3.2 Binary

Since the base of the binary system is 2, it follows that, starting from the units column (the least significant position in an integer value) and working to the left, each place value will be two times greater than its predecessor. Thus the place values are 1, 2, 4, 8, 16, 32, 64, etc, increasing far more slowly than with decimal values which reach the order of millions when the seventh position is reached.

Hence, in the example shown in Figure 3.2 the binary quantity 110101 is seen to have the equivalent decimal value of 53. If, therefore, the number 53 had to be stored in a numeric form in a

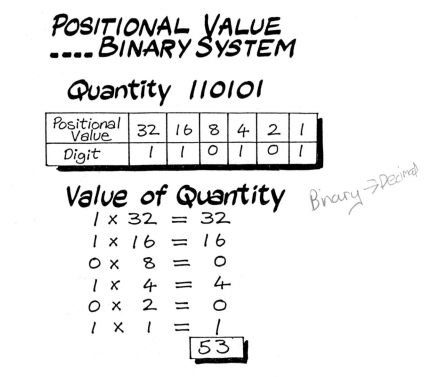

Figure 3.2   **Example of Binary System Values**

computer it would be stored as 110101. Both are alternative forms of representing the same quantity, one being in a manner suited to humans and the other in a form better suited to a computer.

Using the same approach the decimal value of 101010 would be 1×32 plus 0×16 plus 1×8 plus 0×4 plus 1×2 plus 0×1, or $32 + 8 + 2$, or 42.

3.3.3 Octal

The octal number system uses 8 as its base and so the place values moving left from the least significant position are 1, 8, 64, 512, 4096, etc. These values increase far more rapidly than with the binary system, if not so quickly as with decimal. The example given in

Figure 3.3 shows that the value of the octal quantity 3056 is decimal 1582 and, again, these two are simply alternative forms of representing the same quantity.

The decimal value of octal 47023 is given, for example, by 4 × 4096 plus 7 × 512 plus 0 × 64 plus 2 × 8 plus 3 × 1, or 16384 + 3584 + 16 + 3, or 19987.

3.3.4 Hexadecimal

In principle there is no difference in the technique necessary for dealing with hexadecimal quantities. However, because such quantities use the digits A to F as well as the more common 0 to 9, lack of familiarity hampers their mastery. The place values are, from the right, 1, 16, 256, 4096, 65536, etc, increasing at a greater

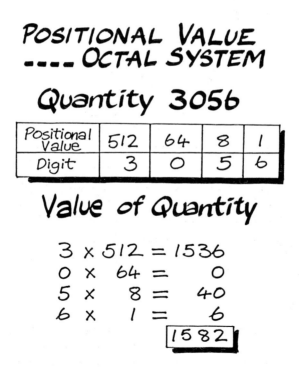

Figure 3.3 Example of Octal System Values

rate than decimal values. In Figure 3.4 the hexadecimal quantity 2FA6 is converted to decimal, yielding a value of 12198.

Using the same method as before the hexadecimal quantity ABC has a decimal value of 10 × 256 plus 11 × 16 plus 12 × 1, or 2560 + 176 + 12, or 2748.

It is now rapidly becoming clear that unless the context makes it absolutely certain, the base being used should be stated when referring to the value of some quantity since, for example, octal 73, decimal 73 and hexadecimal 73 may all appear to be the same yet they have totally different values (59, 73 and 115 in decimal respectively). Generally quantities such as used in the first two chapters of this book are assumed to be decimal. Care must always be exercised however, since it is all too easy to overlook the base used.

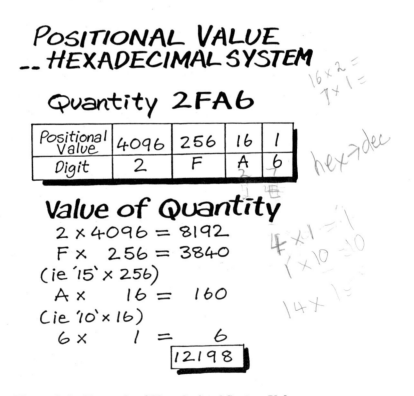

Figure 3.4   Example of Hexadecimal System Values

3.3.5 Fractional Quantities

All that has been said so far relates to integer values but the approach can very easily be extended to deal with fractions as well. Just as in the case of decimal numbers, place values *after* the decimal point refer to units of 0.1, then 0.01, 0.001, etc but moving to the right of the decimal point, and *decreasing* by a factor equal to the base of the number system with each step. On this basis the decimal quantity 36.528 has a value identified by 3 × 10 plus 6 × 1 plus 5 × 0.1 plus 2 × 0.01 plus 8 × 0.001.

In the same manner the binary fraction place values decrease by a factor of two with each step to the right of the bicimal point and are 0.5, 0.25, 0.125, etc. Thus the binary quantity 101.011 has a decimal value given by 1 × 4 plus 0 × 2 plus 1 × 1 plus 0 × 0.5 plus 1 × 0.25 plus 1 × 0.125 or 4 + 1 + 0.25 + 0.125 or 5.375.

With fractions in octal, place values go down by a factor of eight, giving 0.125, 0.015625, etc. Hence the octal quantity 31.27 has a value given by 3 × 8 plus 1 × 1 plus 2 × 0.125 plus 7 × 0.015625 or 24 + 1 + 0.25 + 0.109375 which is 25.359375 in decimal.

Finally, hexadecimal fractions involve us with place values decreasing by a factor of 16, yielding 0.0625, 0.00390625, etc. Thus hexadecimal .CF has a value of 12 × 0.0625 plus 15 × 0.00390625 which is 0.75 + 0.05859375 or 0.80859375 decimal.

3.4 CONVERSION TO DECIMAL FROM OTHER BASES

The preceding sections have established the positional values associated with the most commonly used number bases in computing and, in so doing, have laid down the methods for converting from these bases into decimal values. The following set of exercises provides the opportunity to use these methods.

1. Convert each of the following binary quantities into decimal:

(i)	110		(vi)	110010
(ii)	1001		(vii)	11010
(iii)	11101		(viii)	1110011
(iv)	1101		(ix)	11001101
(v)	10011		(x)	10010110

2. Convert each of the following octal quantities to decimal:

(i)	17	(vi)	362
(ii)	25	(vii)	407
(iii)	43	(viii)	713
(iv)	156	(ix)	1536
(v)	204	(x)	2715

3. Convert each of the following hexadecimal quantities into decimal:

(i)	14 =20	(vi)	A2
(ii)	23 35	(vii)	AB3
(iii)	1B 27	(viii)	B05
(iv)	3C	(ix)	FAD
(v)	7D	(x)	F23E

4. Convert each of the following fractions to decimal fractions:

(i)	1.01 binary	(v)	5.07 octal
(ii)	110.101 binary	(vi)	10.35 octal
(iii)	0.0011 binary	(vii)	3.B hexadecimal
(iv)	2.23 octal	(viii)	19.25 hexadecimal

3.5 CONVERSION FROM DECIMAL TO OTHER BASES

3.5.1 Decimal to Binary

In the conversion from decimal to any other number base the secret is simply to divide the base into the decimal number, note the remainder, then divide the base into the quotient and keep repeating this process until there is a zero quotient which must always occur sooner or later. Reading off the remainders in the reverse order of them being written down will always produce the required answer.

The example in Figure 3.5 shows how this works with the decimal 117 being converted into binary. Since the base of binary numbers is 2, we divide 2 into 117 to get the quotient of 58 and a remainder of 1.

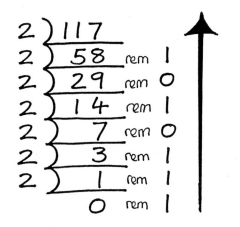

117 Decimal $= 110101$ Binary

Figure 3.5   Conversion From Decimal to Binary

The 58 is now divided in its turn by 2, giving a quotient of 29 and a remainder of 0. The process is continued until at last 2 is divided into 1 giving a zero quotient and a remainder of 1 at which point the process terminates. Reading off the remainders from the bottom upwards gives the answer 1110101 as the binary value of the decimal 117.

3.5.2   Decimal to Octal

In the previous section the general method for conversion from decimal to any other base involved repeated division by the base itself. Hence, if faced with conversion of a decimal quantity into octal we repeatedly divide by 8. In the example given in Figure 3.6, the decimal 236 is converted to an octal value. Division by 8 first produces a quotient of 29 with a remainder of 4. Dividing 8 into this quotient of 29 yields a new quotient of 3 with a remainder of 5 and then the final step is the division of the quotient of 3 by 8 giving zero quotient and a remainder of 3. The process is terminated when a zero quotient is reached and the remainders are read off from the bottom upwards to give the result of 354 octal being the desired equivalent of the decimal 236.

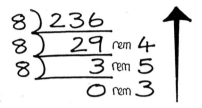

$$236 \; Decimal = 354 \; Octal$$

Figure 3.6   Conversion From Decimal to Octal

Note that the process can only produce remainders of 0 to 7 inclusive so the digits 8 and 9 will never appear in the octal quantity.

3.5.3   Decimal to Hexadecimal

Predictably you should divide by 16 this time, remembering only that if the remainder is between 10 and 15 inclusive, (decimal) it has to be written as A, B, C, etc in hexadecimal.

The example in Figure 3.7 illustrates the use of this technique in converting from decimal 473 into hexadecimal. Division by 16 first produces a quotient of 29 and a remainder of 9. Subsequent division of 16 into the 29 yields a quotient of 1 and a remainder of 13 in decimal terms so D in hexadecimal. The next division produces the

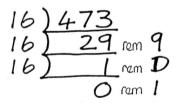

$$473 \; Decimal = 1D9 \; Hexadecimal$$

Figure 3.7   Conversion From Decimal to Hexadecimal

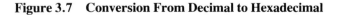

terminating zero quotient and a remainder of 1. Reading the remainders, as before, from the bottom upwards yields the value 1D9 hexadecimal.

Since dividing by 16 can leave remainders (in decimal terms) up to 15 it is no surprise to see the use of A to F as the hexadecimal equivalents of the decimal 10 to 15 appear frequently as remainders.

This particular conversion is no more difficult than from decimal into either binary or octal; lack of familiarity with the use of the 'digits' A to F however makes it appear so. The next exercise gives practice in the use of these conversions.

3.6 EXERCISES IN CONVERSION FROM DECIMAL

1. Convert each of the following decimal values into binary:

(i)	13	(vii)	103
(ii)	25	(viii)	175
(iii)	29	(ix)	236
(iv)	64	(x)	199
(v)	73	(xi)	274
(vi)	127	(xii)	493

2. Convert each of the following decimal values into octal:

(i)	9	(vii)	96
(ii)	17	(viii)	100
(iii)	23	(ix)	127
(iv)	39	(x)	301
(v)	47	(xi)	532
(vi)	83	(xii)	798

3. Convert each of the following decimal values into hexadecimal:

(i)	23	(iv)	42
(ii)	29	(v)	59
(iii)	35	(vi)	79

(vii)	97	(xii)	537
(viii)	140	(xiii)	1295
(ix)	254	(xiv)	1998
(x)	260	(xv)	2748
(xi)	509	(xvi)	12047

3.7   CONVERSION FROM DECIMAL FRACTIONS TO BINARY FRACTIONS

Earlier the positional values associated with binary, octal and hexadecimal fractions were examined and some conversions from these forms into decimal fractions were done; in every case an *exact* conversion was possible. However experience shows that it is not always possible to represent some fractions exactly as decimal fractions; $1/4$ is indeed exactly 0.25 but $1/3$ is 0.3333 recurring and so has no exact decimal fraction equivalent. In precisely the same way some decimal fractions cannot be exactly recorded as binary fractions (nor as octal or hexadecimal) even if they might be exactly reproduced as decimal fractions. An obvious example must be $1/10$ which is exactly 0.1 in decimal terms but can only occur as a recurring binary fraction. This is a major factor in errors produced by a computer when calculating with fractional values, in that it can only represent some of these in an approximate manner, leading to small but accumulating errors when doing calculations with them. Some decimal fractions can be converted exactly and, where this is not possible, the accuracy to which the conversions are calculated can be controlled. The example in Figure 3.8 illustrates the technique employed in converting the decimal fraction 0.743 into a binary fraction. The process involves repeatedly *multiplying* by 2 that part of the decimal fraction which lies to the *right* of the decimal point, writing down the whole number part of the product at each stage (but not involving it in subsequent multiplication). Reading *down from the top* the whole number parts give the binary fraction to as many bicimal places as are required.

Hence when 0.743 is multiplied by 2 to give 1.486, the whole 1 is written down (it will form a part of the bicimal value subsequently) but it is only the 0.486 which is then doubled to give 0.972 at the next

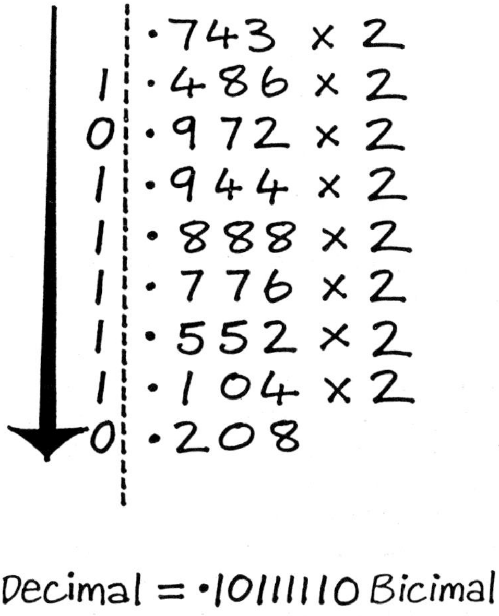

$$
\begin{array}{r|l}
 & \cdot 743 \times 2 \\
1 & \cdot 486 \times 2 \\
0 & \cdot 972 \times 2 \\
1 & \cdot 944 \times 2 \\
1 & \cdot 888 \times 2 \\
1 & \cdot 776 \times 2 \\
1 & \cdot 552 \times 2 \\
1 & \cdot 104 \times 2 \\
0 & \cdot 208 \\
\end{array}
$$

$\cdot 743$ Decimal $= \cdot 10111110$ Bicimal

Figure 3.8   Conversion From Decimal Fractions to Bicimal

stage (note the necessity for recording the 0); this process is repeated eight times to produce a value of 0.10111110 truncated to eight bicimal places.

The bicimal value so obtained can be converted back into a decimal value and yields 0.7421875 clearly indicating that it contains an error effective at the third decimal place if the value is truncated after only so few as eight bicimal places. Hence to reduce errors of this type the computer needs to store such converted values to a large number of binary places; eight, in this case, is clearly insufficient. Had the decimal value requiring conversion been the mixed number 13.743 then the integral and the fractional parts could have been converted separately to yield the binary 1101.10111110.

The conversion from decimal to octal or hexadecimal fractions follows the same process as defined for binary fractions but using multipliers of 8 and 16 respectively.

3.8 INTERRELATIONSHIP BETWEEN BINARY, OCTAL AND HEXADECIMAL SYSTEMS

It will not have escaped notice that the bases of the three systems, 2, 8 and 16 are closely interconnected since $8 = 2^3$ and $16 = 2^4$. This enables us to convert directly between binary and octal and between binary and hexadecimal without the need to use decimal values at all. This facility is especially useful since when writing in machine code it is easier to use octal or hexadecimal which convert easily into the binary that the computer uses without the need for further translation, than to write always in the more laborious binary with the attendant risk of errors creeping in. Dumps from memory are often supplied in octal or hexadecimal format for ease of reading or interpreting even though the machine actually stores instructions and data in binary.

3.9 CONVERSION BETWEEN SYSTEMS

3.9.1 Binary to Octal

The conversions of the first eight binary numbers into octal 0 to 7 need to be memorised and appear in Figure 3.9. The binary quantity is split up into groups of three from the bicimal point outwards, to the left in the case of an integer or to the right for a fractional value (or both in the case of a mixed number). Then the octal values are simply read off to replace each group of three, hence 010 is replaced by 2, 101 by 5 and so on. In cases where a full group of three does not occur naturally then zeroes are added to make them up to three; hence in the second example in Figure 3.10 the solitary 1 is treated

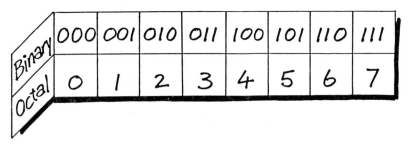

Figure 3.9   Conversion of Binary to Octal

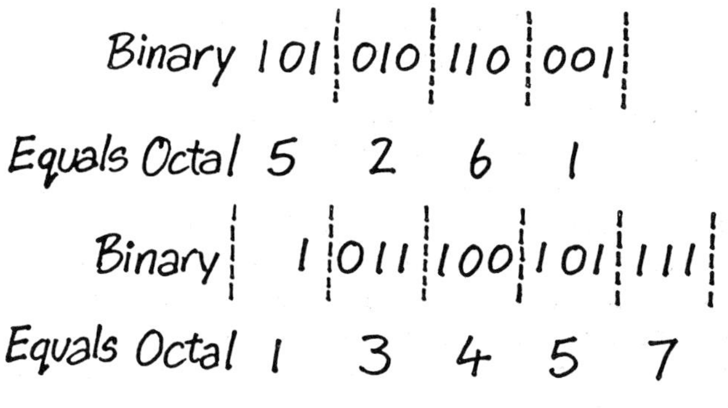

$$Binary\ |0|\ |0|0\ ||0\ |00||$$

$$Equals\ Octal\quad 5\quad 2\quad 6\quad 1$$

$$Binary\ |\ |\ 0||\ |00\ |0|\ |||$$

$$Equals\ Octal\quad 1\quad 3\quad 4\quad 5\quad 7$$

Figure 3.10   Binary and Octal Values

as 001 for conversion purposes. The zeroes if needed are added to the left of integers and to the right of fractional values.

3.9.2   Octal to Binary

This is the exact reverse of the previous case. Each octal digit is replaced by the appropriate 'triple' of binary digits. Any unnecessary zeroes can be pruned afterwards as shown in Figure 3.11.

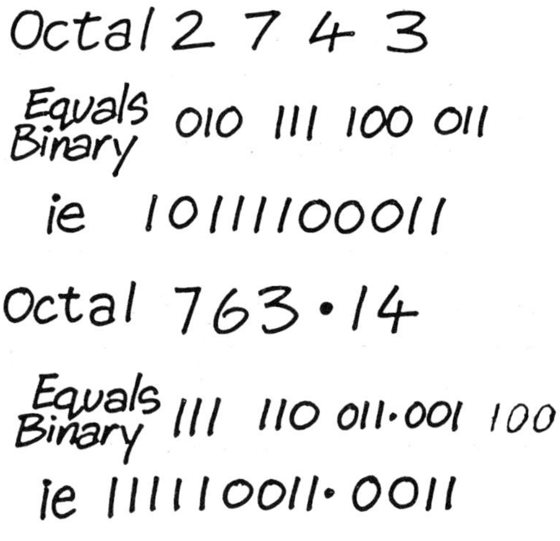

$$Octal\ 2\ 7\ 4\ 3$$

$$\begin{matrix}Equals \\ Binary\end{matrix}\quad 010\ 111\ 100\ 011$$

$$ie\ 101111100011$$

$$Octal\ 763 \cdot 14$$

$$\begin{matrix}Equals \\ Binary\end{matrix}\ 111\ 110\ 011 \cdot 001\ 100$$

$$ie\ 111110011 \cdot 0011$$

Figure 3.11   Conversion From Octal to Binary

Binary	Hexadecimal	Binary	Hexadecimal
0000	0	1000	8
0001	1	1001	9
0010	2	1010	A
0011	3	1011	B
0100	4	1100	C
0101	5	1101	D
0110	6	1110	E
0111	7	1111	F

Figure 3.12　Conversion Table From Binary to Hexadecimal

3.9.3　Binary to Hexadecimal

Once again the conversions between the 16 hexadecimal digits 0 to F must be memorised and these appear in Figure 3.12.

Just as with binary to octal the quantity is split up into groups though this time into blocks of four in binary and the hexadecimal values read off directly. Just as before, zeroes may be added if necessary; the 110 in Figure 3.13 is being treated as though it were 0110.

Binary 110 1011 1011

Equals
Hexadecimal　6　B　B

Figure 3.13　Binary to Hexadecimal Conversion

3.9.4   Hexadecimal to Binary

Predictably this reverses the previous case and a number of examples appear in Figure 3.14.

Hexadecimal   9 AC

Equals Binary   100110101100

Hexadecimal   70 F2

Equals Binary   0111000011110010

ie 1110000111110010

Hexadecimal   62·9E

Equals Binary   01100010·10011110

ie 1100010·1001111

Figure 3.14   Examples of Hexadecimal to Binary Conversion

3.10 EXERCISES

The following exercises test the techniques developed in the past few sections.

1.   Truncating each time after eight binary places, convert each of the following decimal fractions to binary fractions:

(i)   0.528

(ii)   0.9074

(iii)   0.2935

2. Truncating after six octal places, convert 0.392 decimal into an octal fraction.

3. Truncating after five hexadecimal places, convert 0.203 decimal into a hexadecimal fraction.

4. Convert each of the following binary quantities into octal:

(i)	101101110	(v)	0.110101
(ii)	110001100	(vi)	0.1011
(iii)	11011	(vii)	11.01
(iv)	10110	(viii)	1011.1011

5. Convert each of the following octal quantities into binary:

(i)	725	(iii)	23.4
(ii)	536	(iv)	106.53

6. Convert each of the following binary quantities into hexadecimal:

(i)	11010011	(iii)	0.111
(ii)	100101	(iv)	101.01

7. Convert each of the following hexadecimal quantities into binary:

(i)	2B	(iii)	8.C
(ii)	FC7	(iv)	3D.E2

3.11 INTRODUCTION TO COMPUTER-BASED CALCULATIONS

Computers record numbers as binary quantities; so to understand how they carry out calculations requires an understanding of how binary calculations are handled. There are, in fact, no essential differences between the ways in which decimal calculations are handled and those involving binary numbers that are not covered by the differences (and the conversions) between the two number systems discussed in the previous sections, hence there is very little by way of new techniques to be learned. There is, however, a warning that must be given in the context of applying arithmetic in binary to computer calculations. The computer holds several different things in memory, numeric data, character strings, program instructions, all of which are in binary, so that there is no superficial way of telling which of the three is represented for example by 01101101. Even if it is known to be an item of numeric data it still might be integral, fractional or a mixed number or it might be in one of a number of other forms to be dealt with later. Hence caution must be exercised so that calculations on data which is really, for example, representing character strings are not carried out. Although it may be possible to perform the operations the results will be nonsensical and the danger exists that their validity will be accepted.

This problem is less common with high level programming languages than with assembly codes. This book is not the place to discuss how to control this problem but it is well worth pointing out the possibility of it occurring.

Another problem is that different computers carry out certain processes differently from one another though this should not materially affect the work done in this chapter. Addition is, however, generally controlled by hardware and the same addition circuits can be used to deal with subtraction if the numbers are appropriately stored. Multiplication often uses addition circuits coupled with a technique known as shifting. To control these functions requires software which, if permanently embedded in ROM is called *firmware*.

One important limitation is, of course, the fact that computer storage units, be they bytes or words, are of finite size so difficulties may be experienced when storing (and calculating with) quantities too large or too small to be contained in these units. This creates problems involving truncation and rounding of values and the whole area of accuracy is one dealt with later in Chapters 19 and 20.

3.12 ADDITION OF TWO OR MORE BINARY NUMBERS

In order to perform addition in binary we need to be familiar with the following 'rules' of binary addition:

0 plus 0 produces 0 with a 'carry' of 0,

0 plus 1 produces 1 with a 'carry' of 0,

1 plus 1 produces 0 with a 'carry' of 1,

1 plus 1 plus 1 produces 1 with a 'carry' of 1.

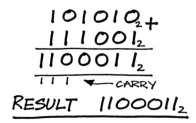

Figure 3.15a   Addition of Two Binary Numbers

The examples in Figures 3.15a and 3.15b show binary numbers added together using these rules, with the 'carry' shown whenever it is non-zero. Note the use of the subscript 2 to denote a binary quantity in Figure 3.15a. If in any doubt as to the result of this sum, one method of checking is to convert each of the numbers being added into decimal (giving 42 and 57 respectively) and then do the same for their 'sum' (giving 99) and confirm that the result is correct by reference to decimal addition.

Needless to say, whilst this provides reassurance as well as more practice at binary to decimal conversions it would be a total waste of effort to do this every time a binary addition was being performed. One point that certainly should be noted is that 42 added to 57 will yield a result of 99 no matter whether the numbers are represented and added in binary as here or in octal or indeed in any other number base whatsoever.

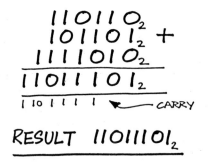

Figure 3.15b   Addition of Three Binary Numbers

In the next example (Figure 3.15b) there are three numbers being added together with the need for adding 1 and 1 and 1 and 1; the result is 0 with a carry of 10 (binary). The decimal values of the three numbers being added are 54, 45 and 122 and that of their 'sum' is 221, providing a check that the result is correct. The 'carry' is shown only when it is non-zero for clarity.

These two examples show the main features of binary addition; if confronted with finding the result of adding up a set of binary digits, it may be best to mentally convert them into decimal and change the total back into binary whilst familiarity is being gained. Hence if faced with $1 + 1 + 1 + 1 + 1 + 1$, which produces decimal 6, or binary 110, the result is 0 with the 'carry' as 11.

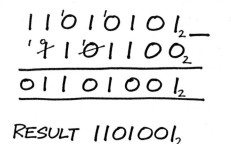

$$1\ 1\ 0\ 1\ 0\ 1\ 0\ 1_2\ -$$
$$7\ 1\ 0\ 1\ 1\ 0\ 0_2$$
$$0\ 1\ 1\ 0\ 1\ 0\ 0\ 1_2$$

$$\text{RESULT}\quad 1\ 1\ 0\ 1\ 0\ 0\ 1_2$$

Figure 3.16   Subtraction of Two Binary Numbers

3.13 SUBTRACTION OF TWO BINARY NUMBERS

The technique for subtracting two binary numbers is quite straightforward but it should be stressed that it is *not* generally the way used for such subtraction in the computer. The computer method, involving the use of complementary arithmetic, is covered fully in Chapter 4. The example shown in Figure 3.16 illustrates the use of 'borrowing' from the next column to the left, a standard approach to subtraction but one which is probably found to be more awkward when dealing with binary quantities. Again, a check is possible since we are subtracting decimal 108 from 213 and the result is indeed the binary representation of 105.

3.14   EXERCISES INVOLVING ADDITION AND SUBTRACTION

Using binary arithmetic calculate each of the following:

(i) $1101_2 + 10111_2$

(ii) $10110_2 + 11011_2$

(iii) $10111_2 + 10101_2 + 110_2$

(iv) $110011_2 + 110101_2$

(v) $1101_2 + 1011_2 + 10110_2$

(vi) $10111_2 + 11101_2 + 10001_2$

(vii) $1011101_2 + 1101011_2 + 110101_2$

(viii) $101101_2 - 10110_2$

(ix) $1011011_2 - 110111_2$

(x) $10101011_2 - 1111101_2$

4 Simple Fixed-length Calculations in the Computer

4.1 REPRESENTING NUMBERS AS FIXED-LENGTH COMPUTER WORDS

In Chapter 3 we introduced what might be referred to as 'pure' binary calculation, along with additional ways in which such calculation might be conducted in the computer. Other differences also exist between 'pure' and computer calculations.

Within a computer numbers are held in fixed-length storage locations (bytes or words) and, in general, we store one number into each location; this at once means that there is an upper limit to how large a number may be stored. For example, suppose that the word size is of 12 bits, despite the possibility of encountering 24- or 32-bit storage locations (there is no difference in principle). When a computer is described as having a byte structure, these bytes will be combined for numeric storage purposes to provide for 8- or 16-bit locations in the same way, so that there is no difference between the use of words and bytes in this respect.

4.2 STORAGE METHODS

4.2.1 Storage of Integers

If we take the decimal number 735 as our first example we are already able (see Chapter 3) to convert this into binary, producing 1011011111, a 10-bit quantity. Since a 12-bit computer word is being used two zeroes will need to be added to the left-hand end of the 10-bit representation in order to store it in this word. Actually, if we look at the 12-bit result, the first (left-hand) bit is not concerned with the *absolute value* of the number but only with its sign (if 0, the number is positive), thus we have only 11 bits left in which to hold the absolute value of the number. The storage of negative numbers is dealt with later in this chapter but it will be seen that in such cases the sign-bit will be 1.

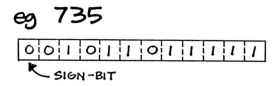

The second example involves storing the decimal value 174; this in 'pure' binary is 10101110, taking up only eight bits, but to store it into the 12-bit word requires us to add four extra zeroes to the left-hand end, including the sign of the number.

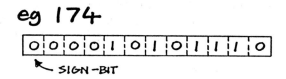

Clearly therefore if the number is quite small more space has to be used up to store the number than actually required; this may well appear wasteful of space but it must be regarded as one of the apparently annoying, but actually very necessary facts of computer storage. To store the number decimal 9 for example would require us to store it as 000000001001.

There is also an upper limit to the size of the number. The largest binary value which can be stored in a 12-bit word is 011111111111 and this in decimal terms is 2047, a very modest upper limit for storage. One commonly-used way of overcoming the problem is to combine two (or more) words together. If this is done an apparently 24-bit word is available for storage. In fact, since the first bit is still the sign-bit and the 13th bit is the sign-bit from the 'second' word (and is not used), only 22 bits are available in which to hold the absolute value of the number. However the value of the largest number is increased to 4, 194, 303 in the double-length word. The example shows 7649 held in this way:

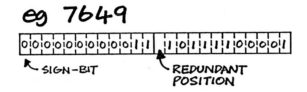

It is possible to combine together as many 12-bit words as may be necessary to hold exact integral values of any magnitude although some accuracy is frequently sacrificed by the use of floating-point representation (see Chapter 5).

With different word lengths the sizes of numbers that can be stored in them will clearly be different from those identified using a 12-bit pattern; for example an 8-bit word can only hold up to 127, but it is still possible to combine together two or more such words to extend the range. Do note however, that if three words are combined in this way then there will be *two* redundant sign-bits in the result.

4.2.2 Storage of Fractions

The first bit still serves as the sign-bit but if a 'pure' binary fraction is smaller than the storage space available in the 12-bit word, then zeroes are added as with integers, but at the right-hand end to fill up the space. Notice that the bicimal point itself is not stored but only implied in the stored quantity; without this factor the binary pattern could just as easily be that of an integer. The example below shows the storage of decimal 0.4140625 which has the *exact* bicimal representation of 0.0110101 and so needs four zeroes to be added at the right-hand end.

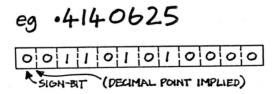

If however the binary representation exceeds the 11 bits available for storage then another choice which is open is to either round or to truncate the result to make it fit the space. This action brings with it the immediate risk of errors caused by making approximations (referred to in Chapter 3 and further discussed in later chapters). This is, in fact, quite common since so many decimal fractions cannot be exactly converted into binary fractions, though still a significant source of potential error. The example below shows decimal 0.7328 stored in a single word, truncating those bits that do not fit into the space available.

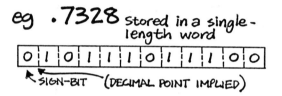

As with integers, the option of combining together two (or more) words is still available; this increases the level of accuracy to which fractional values may be stored giving double-precision working. Once again only 22 bits are available to hold the absolute value of the fraction. In the example below the storage of decimal 0.7328 is shown truncated to fit into a double-length space.

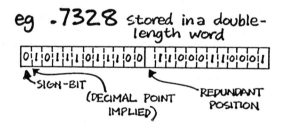

When dealing with fractions the range of values that can be stored in a 12-bit word is not so restrictive as the 0 to 2047 of the positive integers; for all practical purposes the range is from 0 to bicimal 0.011111111111, which is about decimal 0.999512. The use of double-length working does not materially change this range but only provides for more accurate storage of the numbers involved.

4.2.3   Storage of Mixed Numbers

It is possible to store a mixed number in a single word, provided only that an agreed number of bits is allocated beforehand to the holding of the integer part and the balance to the fractional part; in such cases the position of the bicimal point has once again to be implied. Clearly, since we must hold the whole of the integer part, the number of bits left for the fractional part may be so small as to induce errors of relatively large absolute magnitude in the stored quantity. The example below shows the storage of the decimal 11.75; five bits have been allocated for the integer part and six bits for the fraction, hence using 01011 for the integer and 110000 for the fractional part.

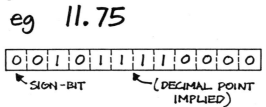

If double-length working is chosen to hold mixed numbers, it is more usual (but not obligatory) to use one word to hold the integral part and for the other to hold the fractional part, making redundant the second (and any subsequent) sign-bit. This method is capable of as much extension as may be desired, allowing for three or even four words to be put together to provide as much accuracy as may be considered appropriate; however, except in quite exceptional situations, double-length is considered quite adequate. The next illustration below shows the storage of the decimal 74.890625 held in two words using the first for the integral 74 and the second for the fractional part.

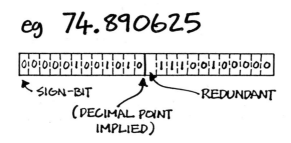

4.2.4   Storage of Negative Values

When distinguishing between +596 and −596 the only way that we indicate that one is positive and the other is negative is by virtue of the + or − sign that is written in front, retaining exactly the same absolute quantity 596 in each case. It is quite easy and, indeed, most obvious that this approach should be extended into the area of computer storage, the approach being called the *sign-and-modulus method*. In this case a 0 is used in the sign-bit for a positive quantity and 1 if it is negative; the rest of the stored quantity remains the same. Hence in the example given in Figure 4.1 the two decimal values +157 and −157 are shown stored using the sign-and-modulus method.

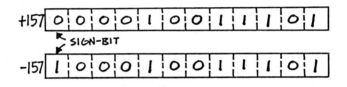

Figure 4.1   The Sign-and-Modulus Storage Method

However, despite the simplicity of this method it does have one outstanding drawback in that it is most unsuitable for the purpose of performing calculations. For this reason the *twos complement method* of storing negative numbers which *is* very well suited to the task of carrying out calculations is examined below.

There are three stages in finding the twos complement representation of a negative number such as that of decimal −837 used in this example. These are:

(i)   Obtain +837 as a 12-bit binary word. Since this involves a positive quantity the method is as described earlier.

(ii)   Invert all 12 bits, that is change every 0 to a 1 and every 1 to a 0.

(iii)   Add 1 to the right-hand, less significant, end.

This is shown, in detail, in Figure 4.2.

Hence the twos complement representation of −837 is 110010111011. The presence of the 1 in the leading (sign-bit) position is a reminder

$$+837\ldots\ldots\ 0 0 1 1 0 1 0 0 0 1 0 1$$
$$(\text{INVERT})\ldots 1 1 0 0 1 0 1 1 1 0 1 0\ +$$
$$(\text{ADD } 1)\ldots\ \underline{\qquad\qquad\qquad 1}$$
$$-837\ldots\ldots\ \underline{1 1 0 0 1 0 1 1 1 0 1 1}$$

Figure 4.2   The Twos Complement Storage Method

that it represents the storage of a negative quantity. This may be slightly more involved than the sign-and-modulus method but it does enable us to use the result for purposes of calculation.

There is an alternative way of considering this representation. Instead of treating the first bit merely as the sign-bit, treat it as having a place value. Just as the place values in binary (from the right to the left) are 1, 2, 4, 8, 16, etc so the twelfth would ordinarily be 2048. Instead give to this 'sign-bit' place holder the value of -2048, whilst according to the other 11 their usual positional values. The following calculation shows quite clearly that converting back from the binary to the decimal, we revert to -837:

$$
\begin{array}{rrr}
1 \times & 1 = & 1 \\
1 \times & 2 = & 2 \\
0 \times & 4 = & 0 \\
1 \times & 8 = & 8 \\
1 \times & 16 = & 16 \\
1 \times & 32 = & 32 \\
0 \times & 64 = & 0 \\
1 \times & 128 = & 128 \\
0 \times & 256 = & 0 \\
0 \times & 512 = & 0 \\
1 \times & 1024 = & 1024 \\
1 \times & -2048 = & \underline{-2048} \\
\text{Value is given by} & & \underline{-837}
\end{array}
$$

Conversely if presented with a 12-bit quantity which was in a twos complement form the process can easily be reversed to convert it back into its decimal value, surprisingly perhaps by again using the three steps used earlier for the twos complement representation of a negative quantity. Given the twos complement representation 100111010110, we proceed as follows:

(i)   Invert all the bits, producing 011000101001.

(ii)   Add 1 to the right-hand end, yielding now 011000101010.

(iii)   Convert this to its decimal value, 1578, and place a minus sign in front, producing the final result −1578.

Hence 100111010110 is the twos complement representation of −1578.

Complements are used in other contexts as well as in computer arithmetic. We might refer to a time of 8.39 as "21 minutes to 9". The 21 and the 39 together add up to 60, the number of minutes in an hour. In this case we might describe 21 as being the complement of 39. Likewise if we were to add together the representations already shown for +837 and for −837 in binary, the result would be the 13-bit quantity 1000000000000; this would be true no matter whether the numbers involved were +368 and −368 or any other such pair of 'complementary' values.

There is also a *ones complement method* for storing negative numbers which is identical to that of the twos complement except that it omits the step of adding 1 at the right-hand end; it is not as widely used as the twos complement method.

When examining the holding of integer values earlier, the range of values available for storage of positive numbers in a 12-bit word was found to be from 0 to +2047 only. The introduction of the twos complement method and the ability to satisfactorily store negative numbers has no effect upon the storage of positive values which remain exactly as before but now the 'most negative' quantity that can be held is 100000000000 or −2048 decimal. Thus the range of integral values that can be held in a 12-bit word is from −2048 up to +2047, a total of 4096 different values. In a similar manner an 8-bit word can hold from −128 up to +127, a total of 256 different values.

Double-length working is just as valuable when dealing with

negative quantities as with positive ones, the process being exactly as already defined, noting again however that the second (and any subsequent) sign-bit remains totally redundant throughout.

4.3 EXERCISES INVOLVING STORAGE METHODS

Using the foregoing techniques do the following tasks.

1. Store decimal 729 in a 12-bit word.

2. Store decimal 97 in an 8-bit word.

3. Store decimal 0.372 in an 8-bit word (truncate as necessary).

4. By combining together two 8-bit words, store decimal 0.372 accurately to 14 bicimal places (truncate as necessary).

5. Using a 12-bit word with seven bits for the integral part, store the decimal value 18.875 exactly.

6. Store decimal −275 using twos complement form in a 12-bit word.

7. A 12-bit word, using twos complement storage format, contains the binary value 110011001100. What decimal value does this represent?

8. Store decimal −57 using twos complement form in an 8-bit word.

9. An 8-bit word, using twos complement storage format, contains the binary value 11000011. What decimal value does this represent?

10. What range of integral values may be stored using a 10-bit word?

11. A 12-bit word, using six bits for the integral part, stores a mixed number as 001011101011. What decimal value does this represent?

12. Store decimal −271 using twos complement form in a 10-bit word.

4.4 BINARY SUBTRACTION USING THE TWOS COMPLEMENT METHOD

If faced with the task of calculating 1382 − 797, the first step in dealing with it is the algebraic one of recognising that this is precisely the same as 1382 + (−797). This, in its turn, shows that in order to subtract 797 from 1382, the complement of 797 (since −797 is the complement of 797) has to be *added* to 1382. From the point of view of computer calculation this is particularly convenient since it means that any subtraction can be carried out by addition, so the addition circuitry can be used for both purposes.

To demonstrate this, consider the calculation of 1382 − 797 shown in Figure 4.3. First find the twos complement form of −797; since +797 is stored as 001100011101, the storage of −797 will be as 110011100011. For the rest, the calculation is now conducted exactly as though it were a straightforward addition; however, given that 12-bit quantities are being used there will always be a carry into a 13th position, at the left-hand end. Clearly, in a 12-bit word there is ordinarily no way such a carry may be held, but this is not important since it *must be totally ignored*. Disregarding this 13th bit, the result is 001001001001 or decimal +585, the correct result of the 'subtraction'.

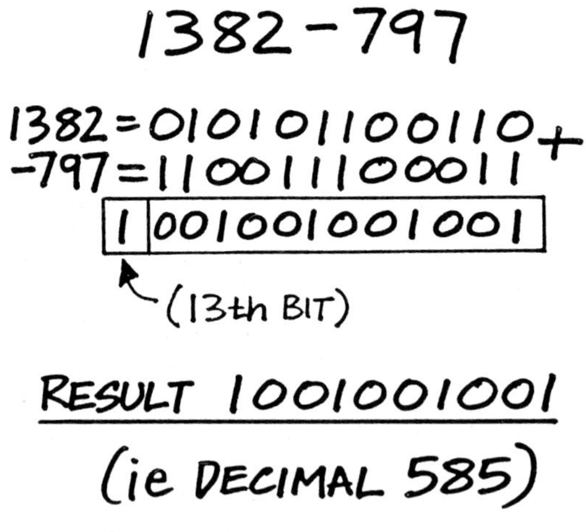

$$1382 - 797$$

$$\begin{aligned}1382 &= 010101100110 \\ -797 &= 110011100011\end{aligned} +$$

$$\boxed{1\ \boxed{001001001001}}$$

$$\uparrow (13\text{th BIT})$$

$$\underline{\text{RESULT } 001001001}$$

$$(\text{ie DECIMAL } 585)$$

Figure 4.3   Twos Complement Subtraction

To illustrate this with another example; consider using 8-bit words to deal with the calculation of 93 − 47. The binary value of 93 is 01011101; that of 47 is 00101111, hence −47 will be 11010001. Carrying out the subtraction yields:

$$01011101$$
$$+$$
$$11010001$$
$$(1)00101110$$

and, ignoring the excess bit at the left-hand end, the result is 101110 or decimal 46 as should have been expected.

4.5   EXERCISES IN TWOS COMPLEMENT SUBTRACTION

Use the method described in the previous section to calculate each of the following:

1. Using 8-bit words,   (i)   87 − 53

    (ii)   107 − 28

2. Using 12-bit words,   (i)   392 − 135

    (ii)   1826 − 399

    (iii)   2217 − 1543

4.6   SHIFT OPERATIONS

Having introduced computer addition and subtraction, it should be recognised that the other two fundamental arithmetic operations, multiplication and division, can be undertaken by processes of repeated addition and subtraction. In decimal terms if we wanted to multiply 59 by 4 we should have to work out 59 plus 59 plus 59 plus 59; to divide 17 by 3 we should have to count how many times we could subtract 3 from 17 until a value of less than 3 was left behind.

Quite clearly then, the earlier processes have a high degree of involvement with those of multiplication and division; however one other operation has a place here, that of shifting. Shifting involves the actual physical movements of bits within a computer word, moving them to the left or to the right. There are a number of different forms of the shift operation such as circular, logical and arithmetic, each with its own special function, but only the arithmetic shift is considered here since it is the one most relevant to the calculation process.

4.6.1   Shift Operations to Achieve Multiplication

First of all consider the decimal number 735; if we move each digit one place to the left, so that each occupies a higher place-value, eg the 7 in the hundreds position moves up into the thousands position, etc this process would leave nothing in the units column, so a 0 would have to be put there. The effect would be to change the number to 7350, thus multiplying the original quantity by 10. If the same thing is done to a binary quantity its value would be increased by a factor of 2 (noting the different bases of the two number systems). If this leftwards shift was done twice the original number would be multiplied by two twice, ie by 2^2 or 4; three places to the left achieves multiplication by 2^3 or 8. In general therefore a shift of n places to the left achieves multiplication by 2^n.

Shifting to the left creates gaps at the right-hand end which must be filled with zeroes, but if dealing with a fixed-length word, there is the risk that, sooner or later, the sign-bit may get changed as other bits move into that position. It is obvious that repeatedly doubling a number cannot alter its sign, so that one of the rules dealing with arithmetic shifting in fixed-length arithmetic requires that the sign-bit does not change during the process if the result is to have arithmetic validity. One way of reducing the risk is to store the number, not in a single-length word, but in a double-length word, reducing the chances that the sign-bit may be changed; but in practice, a special register exists to monitor changes in the sign-bit position and to flag them as and when they occur, thus enabling remedial action to be taken.

The example given in Figure 4.4, using 12-bit words, shows the effect of shifting to the left on the quantity 110010 (decimal value +50).

4.6.2   Shift Operations to Achieve Division

Just as shifting to the left achieves multiplication by the relevant power of 2, so shifting to the right achieves division by the relevant power of 2. This time, as bits get shifted to the right, gaps appear at the left-hand end; the sign-bit does not change at all but the gaps are

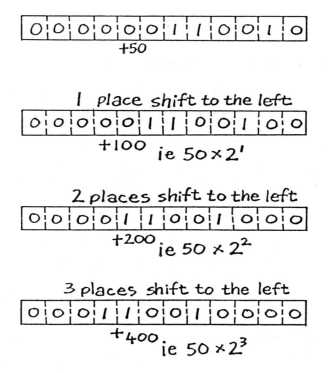

Figure 4.4   Shift Operations to Achieve Multiplication

filled with copies of the sign-bit so that the result retains its arithmetic significance. This is slightly more straightforward than dealing with the comparable situation in shifting to the left. However, bits moving beyond the right-hand end of the word are irrevocably lost leading to a *truncated* form of division (analogous to the idea that decimal 17 divided by 3 is 5). Alternatively if the last bit to be shifted to the right of the units position is recovered and is added back into the bit *now* in the units position, the result will be a *rounded* form of division (analogous to decimal 17 divided by 3 being 6 when rounded to the nearest digit).

In the example given in Figure 4.5, the division of decimal 437 by 8 is achieved by shifting three places to the right, with both truncated and rounded results shown.

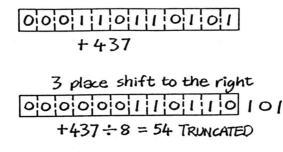

+437

3 place shift to the right

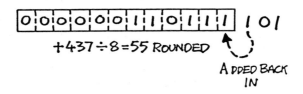

+437 ÷ 8 = 54 TRUNCATED

+437 ÷ 8 = 55 ROUNDED

ADDED BACK
IN

Figure 4.5   Shift Operations to Achieve Division

4.7   EXERCISES INVOLVING SHIFT OPERATIONS

In each of the following cases assume that the quantities are held in 12-bit words.

1. Express decimal 97 in 12-bit binary and calculate the decimal result of shifting four places to the left.

2. To achieve multiplication by 64, how many places must a binary quantity be shifted to the left?

3. Express decimal 537 in binary and calculate the decimal results of shifting three places to the right:

    (i)   using truncated division;

    (ii) using rounded division.

4. Express decimal 1243 in binary and calculate the decimal results of shifting five places to the right:

    (i)   using truncated division;

    (ii) using rounded division.

4.8 BINARY CODED DECIMAL

Sometimes decimal numbers are stored, not as straightforward binary quantities but by coding each decimal digit separately, allocating to each digit four binary spaces (one nibble) and representing each by its 'straight' 4-bit code so that 6 is coded as 0110, 9 as 1001 and so on. This method is called Binary Coded Decimal (BCD) or 'packed decimal'. Whilst it is not such an efficient method of numeric storage as 'pure' binary, it is more convenient for storing very long numbers than representing each in character format, when each would be allocated a whole byte of storage space.

Although not the most convenient form for purposes of calculation, it is at least possible, and the example in Figure 4.6 illustrates how decimal 129 and 346 can be added together in BCD. Note that the corresponding groups of 4 bits are added together and the result is left if a valid BCD code is produced (ie if the result is that of a decimal digit in the range 0 to 9). If the result of the addition is *not* a valid BCD code then 0110 is added to produce one which *is* valid and a carry of 1 goes to the column to the left.

This method is most widely used in cases, (as in many commercial

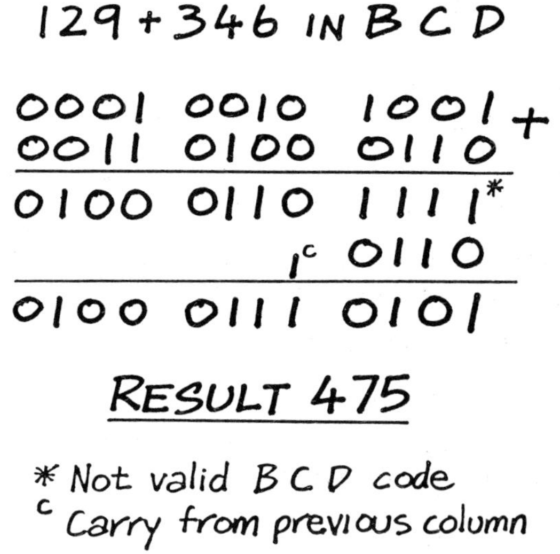

$$129 + 346 \text{ IN } B C D$$

$$
\begin{array}{ccc}
0001 & 0010 & 1001 \\
0011 & 0100 & 0110
\end{array} +
$$

$$
\begin{array}{ccc}
0100 & 0110 & 1111^* \\
& & ^c\ 0110
\end{array}
$$

$$
\begin{array}{ccc}
0100 & 0111 & 0101
\end{array}
$$

RESULT 475

* Not valid B C D code
c Carry from previous column

Figure 4.6 Binary Coded Decimal Calculations

situations), where very long numbers may occur and conversion to pure binary would be very time-consuming (even if done automatically by the computer) but yet calculations are still required.

4.9   EXERCISES IN BINARY CODED DECIMAL

1. Express the decimal 1526 in BCD.

2. By expressing the decimal quantities 253 and 426 in BCD, add them together and convert the BCD result to decimal.

3. By expressing the decimal quantities 1356 and 267 in BCD, add them together and convert the BCD result into decimal.

4.10 FURTHER EXERCISES

1. Store decimal 592 in a 12-bit word.

2. Store decimal −276 in twos complement form using a 12-bit word.

3. Use the results of questions 1 and 2 to calculate 592−276 in binary. Convert your result back to a decimal value.

4. Store decimal 0.735 in a 12-bit word, truncating as necessary.

5. Store decimal 195.427 in a double-length 12-bit word, using one word for the integral part and truncating as necessary.

6. A 12-bit word, using twos complement form, contains 110011101110; what is its decimal value?

7. An 8-bit word, which uses twos complement form for storing negative numbers contains 00110101; what is its decimal value?

8. Express decimal 172 in a 12-bit word and show how it may now be divided by 32, giving your decimal answer in:

    (i)   truncated form;

    (ii) rounded form.

9. Express decimal 37 in a 12-bit word and show how it may now be multiplied by 32 converting your answer to a decimal form.

10. Use BCD to add together the decimal numbers 79 and 536.

5 Floating-point Representation

5.1 INTRODUCTION

We are already aware that, although fixed-length arithmetic is capable of achieving whatever degree of accuracy we might require, it does have a number of limitations. Foremost amongst these is that a single-length 12-bit word is capable only of holding integers in the range -2048 to $+2047$ as previously discussed; if double-length working is used the range of integers that could be held would be increased but only at the cost of taking up additional storage space.

In the same way, there are limitations upon the ranges of decimals and mixed numbers that can be held in a single-length word and whilst increasing the number of words increases the range, it only does so with the penalty of reducing the storage space available.

Whilst the previous discussions have been on the basis of the 12-bit word, exactly the same arguments apply regardless of the size of the word; thus, although a true 24-bit word (*not* the same thing as two 12-bit words because of the redundancy of the second sign-bit) can hold from $-8,388,608$ to $+8,388,607$, it still cannot hold some of the values likely to occur in scientific or commercial arithmetic.

The solution to the problem is a compromise between the need for accuracy and the need to reduce wastage of storage space. Floating-point representation, as the solution is called, allows a far greater range of values, be they integral, fractional or mixed numbers, within the limits imposed by a single-length word. Calculations in floating-point arithmetic tend to be rather slower in execution than in fixed-point working (although they may save the programmer a certain amount of effort, especially with regard to assembly-code programming). One solution to the problem of speed is the installation of special hardware to control and to speed up floating-point calculations even though this may incur a cost penalty. Clearly, if an application is going to involve a great deal of floating-point arithmetic such expense may be worthwhile.

As has already been indicated, the greatest need for floating-point representation occurs when dealing with absolute values that are much larger or smaller than those normally encountered. Examples include the distance from the Earth to the Sun (93 million miles), or the weight of an electron, (0.00000000000000000000000009 grams).

5.2   EXPONENT FORM

It was in areas of science in particular, where such values referred to first arose, so the first attempt at simplifying their representation became known as scientific notation. In this approach, each quantity is expressed as the product of a *mantissa* with an appropriate power of 10 (in decimal quantities), called the *exponent*. (Thus 57429 is expressed as a mantissa 5.7429 times 10^4 when 4 is the exponent.) The absolute value of the mantissa is always greater than or equal to 1 whilst being less than 10. The exponent is always an integer, either positive or negative. Adopting this notation the distance from the Earth to the Sun could be expressed as 9.3×10^7 miles, whilst the weight of the electron could be stated as 9×10^{-27} grams. In the first of these the mantissa is 9.3 and the exponent is 7, whilst in the second the mantissa is 9 and the exponent is -27. There is however, no reason why the mantissa should not also be negative. Hence we might have the value $-287,000,000,000$ to hold in the computer, which in scientific notation would be -2.87×10^{11}, whilst -0.00000000000000528 would be treated as being -5.28×10^{-16}.

5.3   NORMALISED EXPONENTIAL FORM – DECIMAL NOTATION

Unfortunately, scientific notation as it stands, is not the most convenient form for computer storage, because it is very likely that the mantissa will be a mixed number. Therefore an adjusted version is used, called the normalised form, in which the mantissa is effectively divided by 10 and the exponent increased by 1 in compensation.

In this form therefore the mantissa will be a fraction with an absolute value less than 1 but greater than or equal to 0.1, whilst the exponent remains integral. Two of the examples cited earlier (the Earth's distance from the Sun and the weight of an electron) appear in Figure 5.1, showing in each case both mantissa and exponent when using the normalised form; note that it is usual to quote the mantissa with a zero prior to the decimal point, even if this does not subsequently affect the storage. Additionally the manner in which such numbers frequently appear in computer input and output can be seen, using mantissa-exponent form and incorporating the letter E before the exponent appears.

Normalised Exponential Form	Mantissa	Exponent	Computer Input/Output Notation
0.93×10^8	0.93	8	0.93E8
0.9×10^{-26}	0.9	−26	0.9E−26

Figure 5.1   Mantissa-exponent Form

In the same way, if consideration is given to the two negative values referred to earlier, then −287,000,000,000 would be stored with a mantissa of −0.287 and an exponent of 12; with computer input-output notation this would become −0.287E12. Likewise −0.000000000000000528 would have a mantissa of −0.528 and an exponent of −15, and be shown as −0.528E−15 on computer input-output.

5.4   EXERCISES IN NORMALISED EXPONENTIAL FORM

For each of the following quantities, express it in a normalised exponential form writing down both the mantissa and the exponent, and the way in which the number would appear in computer input-output:

(i)	7,400,000	(v)	$-8,200,000$
(ii)	2,560,000,000	(vi)	$-55,400,000,000$
(iii)	0.0000628	(vii)	-0.00027
(iv)	0.000000000153	(viii)	-0.000000079

5.5   NORMALISED EXPONENTIAL FORM – BINARY NOTATION

The same methods apply when extending the ideas of normalisation to binary quantities for computer storage, in that the mantissa will be a positive or a negative fraction with an absolute value greater than or equal to binary 0.1 but less than 1, whilst the exponent will be a positive or a negative integral value. It is important to note that since the mantissa may be positive or negative, one bit must be reserved for its sign, although the absolute value of the mantissa is held *without* recourse to twos complement form if negative.

There are however, two ways of storing the exponent. In the first it is held using the twos complement form, in which case its sign can be determined by examination of the sign-bit in the space allocated to the exponent. In the second way, usually called the 'excess 64 code', the value 2^{t-1} is added to the exponent (where t is the number of bits used to store the exponent). In the situation where 7 bits are used to store the exponent, 2^6 or 64 is added to its value (hence the name), which ensures that what is held is always seen as a non-negative 'pure' binary 7-bit quantity.

The example in Figure 5.2 shows the two alternative methods of storing the binary quantity -10110.1011 in a 24-bit word. Note that in each case the sign of the mantissa appears at the beginning but the 16 bits allocated to its absolute value come at the end; this leaves 7 bits for holding the exponent between the mantissa and its sign.

In the first case the exponent (101 or decimal 5) is held as though in twos complement form (as 0000101) but in the second 64 is added to the decimal 5 to give 69 and this is held as 1000101 in the excess 64 code representation.

5.6   RANGES OF VALUES AVAILABLE

Having considered a 24-bit word using 1 bit for the sign of the

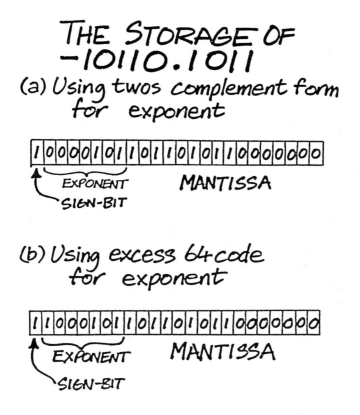

THE STORAGE OF
−10110.1011

(a) Using twos complement form
for exponent

EXPONENT   MANTISSA
SIGN-BIT

(b) Using excess 64 code
for exponent

EXPONENT   MANTISSA
SIGN-BIT

Figure 5.2   Binary Notation in Normalised Exponential Form

mantissa, 16 bits for its absolute value and 7 bits for the exponent consideration should now be given to what ranges of values it is possible to store in such a word using the normalised exponential form.

The 7-bit field allows values for the exponent from −64 to +63, no matter whether the twos complement or the excess 64 code is used. The absolute value of the mantissa is in the range 0.0000000000000 001 to 0.1111111111111111 binary, which in decimal values (to ten decimal places) is from 0.0000152587 up to 0.9999847412.

Concentrating first upon positive values of the mantissa, it is possible to store any decimal number from a smallest value of $0.0000152587 \times 2^{-64}$ (which is exceptionally small) up to a largest value of $0.9999847412 \times 2^{63}$ or $9,223,231,299,000,000,000$.

The numbers with a negative mantissa that can be held in a similar manner have a similar range but with negative values. However, this does mean that a gap exists between the two values $-0.0000152587 \times 2^{-64}$ and $+0.0000152587 \times 2^{-64}$, even if it is an exceptionally small one. Within this gap zero can be denoted by ensuring that the mantissa is itself zero.

This method clearly increases quite dramatically the range of values that can be stored in a 24-bit word from the $-8,388,608$ to $+8,388,607$ range with fixed-point arithmetic, but the penalty which is paid is a restriction upon the accuracy of storage.

5.7 EXERCISES

1. Use the binary version of the normalised exponential form with a 12-bit word, 1 bit for the sign of the mantissa, 6 bits for the absolute value of the mantissa and 5 bits for the exponent to store the decimal values:

   (i) $+59$

   (ii) $+0.35$

   (iii) -39

   (iv) -0.26

   In each case use the twos complement form for storing the exponent.

2. Repeat question 1 using the excess 16 code for storing the exponent (with 5 bits for the exponent, 2^4 or 16 must be added, not 64).

3. Calculate, in decimal terms, the ranges of values that can be stored in a 12-bit word using binary normalised exponential form, given that 1 bit is reserved for the sign of the mantissa, 7 bits for its absolute value and 4 bits for the exponent.

4. Calculate, in decimal terms, the ranges of values that can be stored in a 16-bit word which uses binary normalised exponential

form, given that 1 bit is reserved for the sign of the mantissa, 9 bits for its absolute value and 6 bits for the exponent.

5.8 INTEGER ARITHMETIC

If numbers are held as fixed-point integers then adding, subtracting or multiplying them together will produce correct integer answers, provided only that there is no overflow. However, if attempting to divide two integers it is quite likely that a fraction will be in the answer. If trying to store the result of the division in integer format then truncation (towards zero) will take place so that integer results will occur. Thus you will find that, using integer arithmetic, 15 divided by 7 gives 2, that 9 divided by 11 gives 0 and that -5 divided by 3 gives -1. Some computer programming languages such as Fortran have integer results as a standard; in these cases the fractional part of the quotient is lost and the standard arithmetic rules of division may no longer hold.

5.9 FLOATING-POINT ARITHMETIC

In most practical cases computer calculations are done in floating-point form; the arithmetic associated with this is called floating-point arithmetic or *real* arithmetic. The word 'real' is rather misleading since it implies that anything else is 'unreal' or 'artificial'; in fact it is a word with a special mathematical meaning which is beyond the scope of this book. The rules governing the use of floating-point arithmetic are defined in the following sections.

5.9.1 Floating-point Addition

When dealing with the addition of two numbers which are each expressed in binary normalised exponential form, there are two possibilities; that the exponents are the same or that they are different. This may seem obvious, but each alternative involves using a different method of calculation.

If the exponents are the same then the process merely involves adding the mantissas together and quoting the result using the same exponent as before, unless in the process the addition of the mantissas leads to a value of 1 or greater. In such a case the result has to be normalised, as seen in Figure 5.3 where decimal exponents have been retained solely for convenience.

$$(.1011 \times 2^5) + (.1001 \times 2^5)$$
$$= 1.0100 \times 2^5$$
$$= .1010 \times 2^6$$

Figure 5.3 Floating-point Addition with the Same Exponent

It should be noted that when the two mantissas were added together the result of 1.0100 was greater than that allowed for a mantissa (which must be fractional) so that normalisation had to take place (together with truncation to fit the mantissa into 4 bits).

Where the exponents are different a variation in the technique has to be applied. The quantity with the smaller exponent is adjusted by increasing the exponent to the size of the larger and decreasing the mantissa accordingly (refer to Figure 5.4), which means that the mantissa now has more bits than is normally allowed but at this stage no adjustment is made. Now that the exponents are the same the method described earlier can be used to obtain the result, with normalisation used if necessary and with truncation of the resultant mantissa to enable it to fit into the space allocated as shown in Figure 5.4.

$$(.1001 \times 2^3) + (.1110 \times 2^5)$$
$$= (.001001 \times 2^5) + (.1110 \times 2^5)$$
$$= 1.00000 \times 2^5$$
$$= .1000 \times 2^6$$

Figure 5.4 Floating-point Addition with Different Exponents

In each of the foregoing examples it is quite possible that the mantissa (to which 4 bits were allocated) may exceed the available space and that truncation might take place; this inevitably means that the final result may include some element of approximation, a factor to be considered when discussing the accuracy of the results of such a process.

5.9.2 Floating-point Subtraction

The process is so similar to that of addition that no extra points need to be made. The two examples in Figure 5.5 illustrate the two cases adequately, noting that in the first the presence of non-significant zeroes in the intermediary mantissa (0.0010) require normalisation to be undertaken.

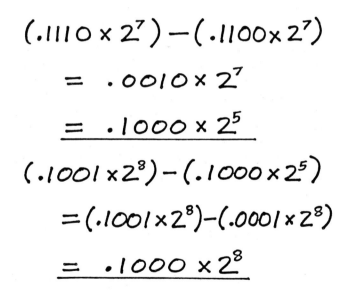

$$(.1110 \times 2^7) - (.1100 \times 2^7)$$

$$= .0010 \times 2^7$$

$$= .1000 \times 2^5$$

$$(.1001 \times 2^8) - (.1000 \times 2^5)$$

$$= (.1001 \times 2^8) - (.0001 \times 2^8)$$

$$= .1000 \times 2^8$$

Figure 5.5   Floating-point Subtraction

5.9.3 Floating-point Multiplication

This is even more straightforward. The mantissas must be multiplied together and the exponents added to give the result. Normalisation and truncation may be needed to produce the final answer. The two

$$(.1101 \times 2^6) \times (.1010 \times 2^4)$$

$$= .1000001 \times 2^{10}$$

$$\underline{= .1000 \times 2^{10}}$$

$$(.1001 \times 2^7) \times (.1111 \times 2^{-11})$$

$$= .10000111 \times 2^{-4}$$

$$\underline{= .1000 \times 2^{-4}}$$

Figure 5.6   Floating-point Multiplication

examples given in Figure 5.6 illustrate the technique but leave the intermediary multiplication of the binary numbers as an exercise for the reader.

5.9.4   Floating-point Division

Once again we have a quite straightforward process. This time divide the mantissas (as far as is necessary to get the correct number of significant digits) and subtract the exponents. Normalisation and truncation may once more be needed to produce the final answer (Figure 5.7).

$$(.1011 \times 2^7) \div (.1101 \times 2^4)$$

$$\underline{= .1101 \times 2^3}$$

$$(.1111 \times 2^8) \div (.1000 \times 2^{-4})$$

$$= 1.111 \times 2^{12}$$

$$\underline{= .1111 \times 2^{13}}$$

Figure 5.7   Floating-point Division

5.10 EXERCISES IN FLOATING-POINT (REAL) ARITHMETIC

1. Using integer arithmetic and decimal quantities, calculate each of the following:

(i) 5×17 (vi) $-4 \div 2$

(ii) $93 + 162$ (vii) $-17 \div 5$

(iii) $493 - 76$ (viii) $28 \div 6$

(iv) $12 \div 3$ (ix) $19 \div 21$

(v) $10 \div 6$ (x) $-22 \div 6$

2. Using floating-point addition calculate each of the following, assuming that the mantissa is allocated 5 bits and that decimal exponents are retained for convenience.

(i) $(0.10110 \times 2^4) + (0.10001 \times 2^4)$

(ii) $(0.10011 \times 2^3) + (0.10010 \times 2^6)$

(iii) $(0.10001 \times 2^2) + (0.11101 \times 2^{-1})$

3. Using floating-point subtraction, calculate each of the following, assuming that the mantissa is allocated 5 bits and that decimal exponents are retained for convenience.

(i) $(0.10110 \times 2^4) - (0.10001 \times 2^4)$

(ii) $(0.10110 \times 2^7) - (0.10101 \times 2^4)$

(iii) $(0.11001 \times 2^3) - (0.10111 \times 2^{-1})$

4. Using floating-point multiplication calculate each of the following, assuming that the mantissa is allocated 4 bits and that decimal exponents are retained for convenience.

(i) $(0.1011 \times 2^3) \times (0.1001 \times 2^4)$

(ii) $(0.1101 \times 2^4) \times (0.1111 \times 2^5)$

5. Using floating-point division, calculate each of the following, assuming that the mantissa is allocated 3 bits and that decimal exponents are retained for convenience.

(i) $(0.111 \times 2^5) \div (0.101 \times 2^3)$

(ii) $(0.110 \times 2^4) \div (0.101 \times 2^{-3})$

6. In the following 12-bit words, numbers have been stored using binary normalised exponential form with the fields appearing in the following order:

    1 bit for the sign of the mantissa;

    5 bits for the exponent (twos complement form);

    6 bits for the absolute value of the mantissa.

    By using floating-point addition add the two quantities together and store the result in the same form.

    The 12-bit words are:

    000110101010

    and   000111100101

6 The Representation of Statistical Data

6.1 THE VALUE OF STATISTICS IN DATA PROCESSING

Within the context of computing it should be realised that one of the functions of any management information system is to accept as input a (potentially) very large volume of individual records and to produce as output a well-organised and coherent summary. To achieve this objective both the tools and the techniques of statistics are needed in order to be able to summarise this raw data and to be able to analyse it and draw inferences from it. This will include, in one company for example, data relating to the volume of sales of a variety of different product lines over a period of time with the need to establish which of these lines are increasing in popularity and which are declining. At the same time the company will need to establish where these lines achieve their best sales figures. Depending upon the results, production levels may have to be adjusted, marketing strategies determined and deliveries to retail outlets varied to suit the patterns which emerge. Similar situations frequently occur and require statistical analytical techniques in order to cope with them.

In the context of the computer department itself there is a need to be able to apply the same techniques to analyse machine downtime, to regulate machine loading and to predict demand for consumable items such as printer ribbons, listing paper or disks. There is also a need to examine the efficiency of software packages, staff productivity levels and the effectiveness of new products for the machine room before being committed to expenditure on their purchase.

At a rather more sophisticated level statistical techniques are needed to assist in the organisation of file and database structures and in 'fine-tuning' to make systems work as efficiently as they are capable of doing.

Since the majority of management information systems are capable of producing screen-based output, the use of graphical representations of data is generally found to create a more easily understood analysis than is the production of tabulated results, although in most such systems *both* have an important place. The ability of a person to absorb graphical output more rapidly than tables of data is a major factor in the design of output, whether to microcomputer screens, to remote VDUs or to printers. The only limitations to the successful development of this are the resolution available on the screens in graphics mode and the level of statistical knowledge of those responsible for designing the systems.

6.2 TYPES OF DATA

At this stage, a definition of the forms that data can take may be useful, as certain types of graphical representation are more suitable than others for some forms of data.

Qualitative data is characteristically not capable of being measured; for example, it includes the use of classification into different colours, types of material, personal preferences in respect of music, books, etc. If a survey were to be carried out to determine the colours of clothing being worn by a group of people the items such as 'red coat', 'blue coat' and 'yellow coat' would, therefore, be described as qualitative. In the same way, the materials such as nylon, aluminium, paper, etc fall into a similar classification.

Whilst we can measure the number of people *wearing* red coats we cannot assign any sort of numerical value to the concept of 'red'; whereas if we consider the number of people with size 7 shoes we *can* assign a numerical value, 7, to the concept of shoe-size. Characteristically we cannot place qualitative data in any particular order uniquely; for example, should red come before yellow or after it?

Quantitative data however, is capable of being put into a well-defined sequence so that shoe size $6^1/2$ comes before size 7, which in turn comes before size $7^1/2$ and so on. It is therefore not possible to assign to qualitative data the ideas of 'greater' or 'lesser' importance since such terms as these, which enable comparisons to be made between two or more data items, can only refer to data which is quantitative. Quantitative data is measurable and therefore includes such things as length, volume, weight, temperature, etc to each of which numerical values can be assigned and hence some idea of ordering can be ascribed to them.

Quantitative data can itself be subdivided into that which is continuous and that which is discrete (Figure 6.1). In the case of *continuous* data every value within a range of values may be found, whereas with *discrete* data only a limited number of values may be found within a given range. Continuous data includes time, length,

TYPES OF DATA

QUALITATIVE:
- Colour
- Types of material
- Foods
- Favourite singers

QUANTITATIVE:

CONTINUOUS:
- Length of a road
- Weight of a parcel
- Age of a building
- Temperature in a room

DISCRETE:
- Shoe size
- Clothing size
- Number of people in a room
- Sizes of tins of food

Figure 6.1 Types of Data

weight, etc since, in theory at least, every value of time between, for example, 6 seconds and 7 seconds can be measured even if the value may be 6.3852187 seconds. In the case of discrete items shoe sizes may be 41, 42, 43, 44, etc but such intermediate values as 41.7 just cannot occur. Paint likewise may be bought by the 0.25 litre, by the 0.5 litre or by the litre but it is not normally possible to buy a tin of 0.3728 litres. Volumes are usually regarded as being continuous but in the specific context of such items as the sizes of tins of paint, it is correct to describe the data as being discrete.

6.3   TABULATION OF DATA

'Raw' data usually consists of large volumes of figures, frequently ill-coordinated, so that the user is swamped by their sheer magnitude. The first stage in making this data usable is to refine it into tabular form, organised into the most convenient order and summarised. Presented below in Figure 6.2 is a list of the number of sheets of continuous listing paper used for each of 120 different jobs run on a computer in a particular day. As it appears at present it is not possible to easily extract much useful information from it.

Number of sheets of listing paper used
on each of 120 jobs

17	8	14	17	5	9	11	18	22	14	6	17
24	11	18	7	21	14	12	27	13	12	9	18
14	29	13	8	9	16	27	21	14	11	19	7
18	14	21	27	11	10	19	14	12	17	9	12
23	16	7	14	21	17	19	24	26	2	5	18
17	24	13	17	8	14	13	28	16	7	8	14
19	16	18	24	7	14	16	19	11	17	23	12
27	9	8	19	13	25	18	21	10	15	11	14
9	8	20	16	8	11	22	10	17	9	18	12
14	28	12	10	9	24	20	5	16	7	10	7

Figure 6.2   List of Data for Tabulation

In order to present this in a more usable form, the number of sheets used has been divided into a number of categories, usually consisting of equally-sized subdivisions such as 5–, 10–, etc. A tally has then been conducted to determine how many jobs come into each of these categories, using a single stroke to indicate each job and the use of a cross-stroke on every fifth job to make counting easier later on. When added up, the total of the strokes is the *frequency* associated with each of the categories. This frequency is the number of times the particular category is to be found in the distribution as a whole, and the resulting table (with or without the tally marks being shown) is called a *frequency distribution* since it shows the frequency associated with *each* category into which the original data was subdivided (Figure 6.3).

Note that the category 10– is to be taken as meaning 10 to 14 inclusive since the data is discrete. The frequency associated with this particular category is 37, indicating that there were 37 entries in the original ill-coordinated table which were in the range 10 to 14 inclusive.

Number of sheets of listing paper used	Tally	Frequency
0–	I	I
5–	︱︱︱︱ ︱︱︱︱ ︱︱︱︱ ︱︱︱︱ ︱︱︱︱ I	26
10–	︱︱︱︱ ︱︱︱︱ ︱︱︱︱ ︱︱︱︱ ︱︱︱︱ ︱︱︱︱ ︱︱︱︱ II	37
15–	︱︱︱︱ ︱︱︱︱ ︱︱︱︱ ︱︱︱︱ ︱︱︱︱ ︱︱︱︱ I	31
20–	︱︱︱︱ ︱︱︱︱ ︱︱︱︱ I	16
25–	︱︱︱︱ IIII	9

Total of all frequencies = 120

Figure 6.3   Frequency Distribution Table

6.4   EXERCISES

1.   For each of the following, decide whether the data is qualitative

or quantitative; if quantitative decide also whether it is discrete or continuous:

(a)   The temperature in the machine room.

(b)   The number of boxes of printer listing paper delivered by lorry.

(c)   The makes of microcomputers used in a company.

(d)   The time taken to execute an instruction in the processor.

(e)   The number of staff employed in the data processing department.

(f)   The number of lines of code in each of a number of programs.

(g)   The size of main memory in Kbytes.

(h)   The functions undertaken by a systems programmer.

(i)   The number of runs of a program to get a 'clean' compilation.

(j)   The salaries paid to staff in the data processing department.

2.   The following is a list of the salaries paid to staff in the data processing department each week. There are 70 staff and all salaries are quoted in $.

314	65	381	174	219	78	156	198	274	315
279	82	186	287	153	158	341	287	164	236
183	176	291	310	264	109	316	192	112	206
194	319	354	264	154	322	213	352	219	341
263	174	207	158	291	371	150	171	286	136
374	163	148	78	273	286	163	253	172	118
286	293	376	107	241	159	88	164	154	209

Using categories of 60−, 80−, 100−, etc up to 380−, conduct a tally and so produce a frequency distribution for the foregoing data.

3. When conducting a benchmark test on each of 50 micro-
    computers, the following times, in seconds, were recorded to
    perform a specified task:

416	289	517	275	496	316	417	316	309	453
374	361	428	538	299	414	586	480	516	347
288	351	500	550	463	512	542	314	531	376
459	308	486	345	406	419	599	398	319	428
374	410	292	290	353	327	417	484	418	572

    Using categories of 275 –, 300 –, 325 –, etc up to 575 –, conduct
    a tally and produce a frequency distribution for the foregoing
    data.

6.5 PICTORIAL REPRESENTATION

For many people, a diagram may convey quickly far more information than can a table of figures; the adage that 'one picture is worth a
thousand words' is certainly appropriate here. A table may contain
a considerable amount of detailed information but it tends to be dull
and to lack impact. On the other hand the diagram, if used effectively,
gives an immediate overview of the contents of the table and can be
more easily digested even if it tends to lack some of the finer details.
It is common, in practice, to produce both diagram *and* table, to
provide the combination of impact, overview and detail that any
user may require.

The following sections examine a number of different types of
diagrammatic representations that can be used in order to produce
the required impact.

6.5.1 Pictograms

If told that in a survey 4000 visual display units each cost between
$600 and $700 we might decide to represent this frequency of 4000
by showing four pictures of a VDU, each of these pictures representing
1000 VDUs. The following diagram extends the use of this idea with
the tabular information appearing at the top and the pictogram next
in Figure 6.4:

Cost of VDU ($)	Frequency
600 –	4000
700 –	5500
800 –	2000
900 – 1000	500

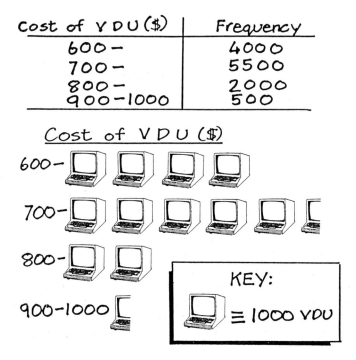

Cost of VDU ($)

KEY:
☐ ≡ 1000 VDU

Figure 6.4   Example of Pictogram Representation

The point of the pictogram, which is alternatively known as an ideograph or as an isotype chart, is that it uses symbols that are easily identified with the item whose frequency is to be represented. Although it is possible to subdivide these symbols, as was indeed seen in the VDU pictogram, it is not advisable to do so beyond the use of half-units since otherwise there are likely to be problems regarding the accuracy of the drawing, as by doing so, it may seem to claim a higher degree of accuracy than it is capable of doing.

Whilst the system is quite versatile (depending largely upon the imagination of the person producing the drawings and despite the fact that it copes equally well with data which is qualitative or quantitative), it is nonetheless open to the risk of abuse.

In the diagram in Figure 6.5 the symbol on the left stands, as before, for 1000 VDUs but the one on the right is twice as high (and

Figure 6.5 Ambiguous Pictogram Representation

in proportion elsewhere). Since it is twice as high it might be regarded as standing for 2000 VDUs; alternatively, it could be argued that since it occupies four times the area on the page (since both height *and* width are doubled) it should be regarded as standing for 4000 VDUs. There is also the argument that, since its volume would be eight times that of the original (height *and* width *and* depth are all doubled) it should stand for 8000 VDUs. To avoid any such controversy, the simplest solution is to decide on one symbol of a given size and to use it throughout the pictogram, with no attempt to use others of different sizes unless they are very clearly identified in the key and their representations unambiguously stated.

6.5.2   Pie Charts

The pie chart uses a circle to represent the whole distribution with individual sectors to represent the individual categories into which the distribution is subdivided. The radius of the circle is not critical so it can be of any convenient size.

The angle at the centre of each sector is directly proportional to the frequency of the quantity represented by that sector; as the area of the sector is *also* proportional to this angle it means that the area too is directly proportional to the frequency. Alternatively since the arc length of the sector is proportional to the angle at the centre it also can be taken as representing the frequency with no risk of confusion. Hence the frequency can be identified directly with either the angle at the centre of the sector *or* with the sectorial area *or* with the arc length of the sector. The diagram shown in Figure 6.6 is illustrative of the way in which such a pie chart is created; note that since the total frequency (for the whole distribution) is 80, this must be proportional to the angle of the whole circle, 360°. Thus

1978 Distribution of staff in Computer Centre

Category	Number Employed
Data Preparation	18
Data Control	5
Operators	18
Programmers	21
Analysts	15
Managers	3

Total = 80

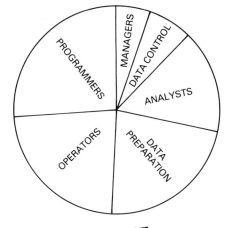

$$\text{Angle for analyst sector} = \frac{15}{80} \times 360° = 67\frac{1}{2}°$$

Figure 6.6   Data Table Converted into a Pie Chart

each person must be represented by a sector angle of $360° \div 80$ or $4^{1}/_{2}°$. For a sector to represent 18 people (the Operators) it must have an angle of 18 times $4^{1}/_{2}°$ or $81°$; the other calculations are made in the same way. Hence the angle for Data Preparation staff must also be $81°$, that for Data Control is $22^{1}/_{2}°$, for Programmers it is $94^{1}/_{2}°$, for Analysts $67^{1}/_{2}°$ and for Managers it is $13^{1}/_{2}°$. Note that these angles should be checked to ensure that they *do* total $360°$.

Generally speaking, pie charts tend to be drawn so that the sectors appear in order of magnitude for visual effect but there is no other motive in doing so. However, occasions do arise when there is a need to have two or more pie charts in the same diagram, usually for the purpose of comparison. For example, the previous pie chart showed the distribution of 80 staff in a data processing department in 1978; a second diagram could perhaps be drawn to show the equivalent breakdown for the same department in 1983 when there were a total of 135 staff. Since the two pie charts will appear side by side it is important to ensure that they *do* achieve the correct visual comparison. The eye is usually more influenced by area than by angle or arc length so to achieve this the areas of the two circles are in the ratio of the staff totals, 80:135. Thus, assuming that the radius of the smaller, 1978, circle had been chosen as 4 cms, the radius of the larger, 1983, circle is calculated as follows:

$$4 \times \sqrt{\frac{135}{80}} = 5.2 \text{ cms (correct to 1 decimal place)}$$

The sizes of the sectors for the 1983 circle are calculated in precisely the same manner as was used earlier for the 1978 chart. The table below provides the necessary data for 1983 with angles calculated in accordance with the rules already given:

Category	Number Employed in 1983	Sector Angle
Data Preparation	25	67°
Data Control	8	21°
Operators	32	85°
Programmers	36	96°
Analysts	31	83°
Managers	3	8°
Total	135	360°

Table 6.1   Distribution of Staff Data for 1983

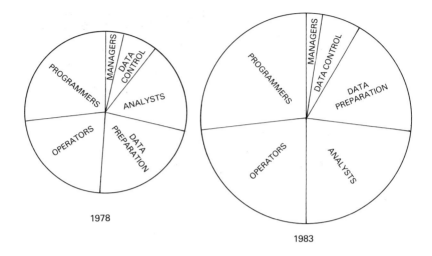

1978

1983

Figure 6.7   Comparison Pie Charts

Figure 6.7 shows the two pie charts, drawn as indicated, enabling full comparisons to be made within each year, as well as between years.

6.5.3   Bar Charts

Bar charts provide an alternative method of representing data that could be represented on a pie chart; hence they are equally suitable for qualitative or quantitative data. The main difference is the use of vertical 'bars' instead of the sectors of a circle. The bars are of equal width so that their areas are directly proportional to their heights, so that the frequency represented by each is given either by height *or* by area. In statistical theory it is the area of the bars which is the more important but, provided that the bars have the same width, the heights may be regarded as measuring the frequencies. The vertical scale (at the left of the chart in Figure 6.8) indicates frequency and is linear, so that equal intervals on the scale denote equal changes of frequency. The bars themselves are set out on a horizontal axis with gaps between them and with annotation to indicate what each bar represents. The bars are commonly arranged to be either in ascending or in descending order of size but can be

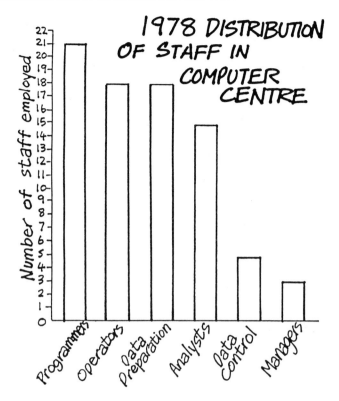

Figure 6.8   Bar Chart Representation

variable. The bar chart in Figure 6.8 shows the same 1978 staff distribution figures previously illustrated using a pie chart.

Sometimes bar charts may be encountered drawn so that the horizontal and the vertical axes are exchanged with the bars horizontal starting from a left-hand vertical base; they are not different in any other way.

There may be situations where the bar chart compares categories the frequencies of which are of the same order of magnitude and to draw them full-scale would present a number of bars all of much the same height. In such cases it is acceptable to produce a break in each of the bars as shown in Figure 6.9; this allows the use of a larger scale

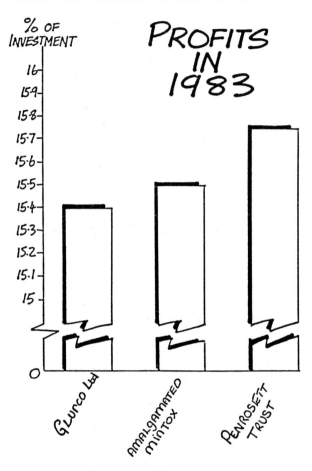

Figure 6.9   Bar Chart with Breaks for Clarity

than might otherwise have been possible, to show the differences between the categories with greater clarity. The practice is quite acceptable but must be shown clearly, as in Figure 6.9, lest it be used to deliberately mislead.

Reverting to the 1978 staff distribution figures we may *also* wish to know the numbers of males and females within each category. In this case, the bars may be subdivided in a manner which ensures their widths are unchanged, producing what is called a Component Bar Chart. There may be more than two subdivisions within each category making a key essential to describe each of the subdivisions and how it is represented. This type of chart allows comparison of the changing composition of the subdivisions within a category and comparison of the subdivisions with one another on a category-by-category basis, in addition to comparison of the categories themselves. The diagram in Figure 6.10 therefore enables us not only to view the male-female division within each of the categories, but also the way in which each sex is distributed within the depart-

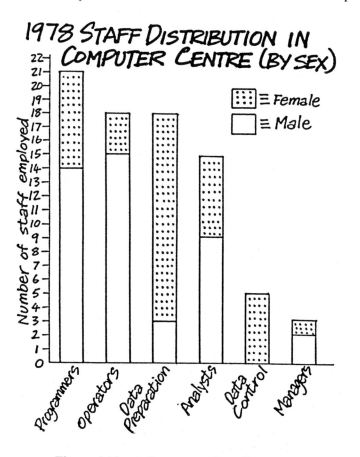

Figure 6.10   A Component Bar Chart

ment in addition to the original, more 'global' purpose of the bar chart.

There are, of course, other variations in the ways in which a bar chart may be presented but the foregoing provide a totally adequate working basis. Do note however, that the horizontal axis as used here does *not* have the same function it might have on a more traditional graph.

6.5.4   Histograms and Frequency Polygons

The methods introduced so far can be used either for qualitative or for quantitative data but those introduced in the rest of the chapter can be applied only to data which is quantitative. Equally well, consideration will be given to whether differences of treatment are necessary for discrete as opposed to continuous data.

The histogram is very like a bar chart in appearance except that the gaps between the bars (which are still of equal width) vanish and the horizontal axis has a linear scale on it. The range of values attributable to each of the categories is shown on the horizontal scale so the order in which the categories are shown depends upon this scale and not upon any ascending or descending order based upon frequencies. The vertical axis continues to represent the frequency attributable to each category.

In Figure 6.11, a histogram has been drawn to illustrate the number of items distributed according to age; there are, for example, some 600 items between 15 and 20 years old whilst there are none in the 0 to 5 year old category.

The data here, identifying the age of equipment (called the *independent variable*) is continuous, so that the categories of 5–10, 10–15, etc share common boundaries, in this case at 10 years, so that 10 years is at the same time the upper limit of one classification and the lower limit of another. It is conventional in such cases to allocate half of those items which are *exactly* 10 years old to each of the categories.

The histogram shown has another feature. The mid-point of the tops of each of the bars have been plotted and joined by a series of straight lines to create what is called a *frequency polygon*. Although

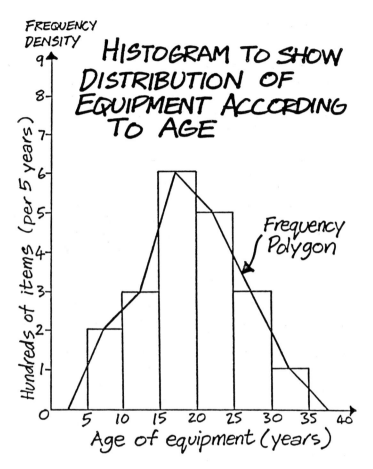

Figure 6.11   Histogram with Frequency Polygon

no bars existed for 0–5 and for 35–40 these, being the categories immediately below and above those shown, are identified with zero frequency and points plotted on them to close off the polygon. If the widths of the bars were reduced to be made very narrow allowing a large number of such bars to exist then the frequency polygon would consist of a very large number of short straight lines and would approximate to a smooth curve, called a *frequency distribution curve*. This curve has considerable importance in statistics and can take one of a number of different characteristic shapes depending upon how the data was originally distributed.

In the next diagram (Figure 6.12) is a histogram associated with discrete data; here the variable (the number of people) being dealt with has to be integral, so that categories of 0–4, 5–9, 10–14, etc are appropriate. However, rather than have a gap between the boundaries at 4 and 5 we establish the convention of a common boundary of $4^1/2$ for the 0–4 and 5–9 categories with equivalent values elsewhere.

The other feature in the diagram is the presence of a category the

Number of people working in each department	Number of departments
0 – 4	5
5 – 9	8
10 – 14	14
15 – 19	17
20 – 24	12
25 – 29	8
30 – 34	6
35 – 49	6

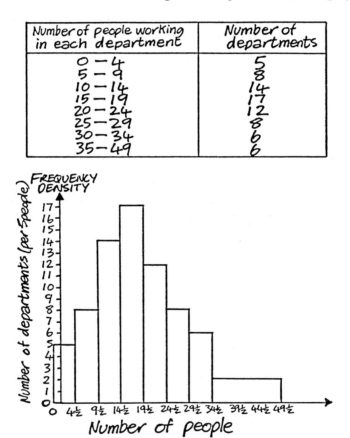

Figure 6.12   Histogram Showing the Importance of Area

width of which is *not* the same as that of the others in the distribution. The category 35–49 is three times as wide as the others. In a histogram it is the *area* of the bar, not its height, which represents its frequency, so since the frequency of the 35–49 category is stated as being 6, the height of the bar must be one third of 6 ie 2, to compensate for the treble width. This accounts for the vertical axis being marked as the frequency density rather than as the frequency, since it really refers to the frequency per '*standard unit* of bar-width'.

6.6   EXERCISES

1.   The following data refers to the number of people employed in computing jobs in a particular town:

Programmers	250
Data Preparation	150
Systems Analysts	175
Operators	300
Managers	25

Create a pictogram to represent the above.

2.   The following list shows the number of computers from each of a number of different manufacturers in a small area of the country:

IBM	12
ICL	7
Unisys	4
Tandem	5
Honeywell	9
Hewlett Packard	7
DEC	14
Data General	2

Represent the above by   (i) a pie chart,

(ii) a bar chart.

3. Create a component bar chart to illustrate the following distribution of jobs undertaken by six companies during a recent month:

Company	Batch jobs	Real-Time jobs	Maintenance
Abloco Ltd	8	24	6
Benserjal Ltd	4	9	1
Colquot Associates	3	17	3
Danvertoft	9	12	6
Eliasbund & Co	16	14	4
Fletchgunders	12	10	7

4. Produce pie charts to illustrate the following analyses of time lost (in hours) in May 1983 and in May 1987:

Month	Industrial Action	Hardware Faults	Software Faults	Human Error	Total
May 1983	6	7	15	3	31
May 1987	2	5	34	8	49

5. Draw both a histogram and on the same diagram a frequency polygon to represent the following data for monthly salaries. (First decide whether the data is continuous or discrete.)

Salary Range ($)	Number of staff
0–100	7
100–200	12
200–300	19
300–400	27
400–500	14
500–600	8
600–700	2

6. Draw both a histogram and on the same diagram a frequency

polygon to represent the following data in relation to the number of runs taken to get a program working for each of a group of 30 programming students:

Number of Runs	Number of students
1–3	4
4–6	6
7–9	9
10–12	4
13–15	3
16–21	4

6.7 CUMULATIVE FREQUENCY DIAGRAMS

Sometimes a frequency distribution as it has so far been presented is not sufficient. The total frequency up to a particular value on the horizontal axis (the value of the relevant variable) may be needed. For example, the table in Figure 6.13 shows an analysis giving a breakdown of the number of lines of Cobol code in each of the programs in the program library file. It shows that 24 programs have between 150 and 174 lines of code; however, we may be interested in knowing how many have less than 175, this being the sum of the

Lines of code	Number of programs
100–	3
125–	12
150–	24
175–	42
200–	51
225–	39
250–	30
275–	21
300–	18
325–349	6

Figure 6.13   Table Showing Breakdown of Number of Lines of Code in Different Programs

Lines of code (less than)	Cumulative frequency
100	0
125	3
150	15
175	39
200	81
225	132
250	171
275	201
300	222
325	234
350	240

Figure 6.14   Cumulative Frequency Table

frequencies in the categories 100–, 125– and 150–, or a *cumulative frequency* of 3 + 12 + 24 = 39.

To do this it will be essential to first construct what is called a *cumulative frequency distribution table* (Figure 6.14) by counting up the total frequency below each of the boundary values given in Figure 6.13, including the lower limit of the first category as well as the upper limit of the last. Hence, the total frequencies for values less than 100, 125, 150, etc need to be found. Since there is nothing less than 100 the cumulative frequency 'less than 100' must be 0; for 'less than 125' it is 3, for 'less than 150' it must be 3 + 12 or 15 and so on, with the final cumulative frequency 'less than 350' being the total of all the original frequencies, ie 240, a check that the calculation has been correctly done.

The table thus shows that 15 programs had less than 150 lines of code each, that 201 had less than 275 lines of code each, etc. To produce the corresponding cumulative frequency curve the values obtained from the cumulative frequency table are plotted. The cumulative frequency (or cf) of 15 is positioned against the 'x' value of 150, 201 for the cf against an 'x' value of 275 and so on, until all the values have been plotted. Once the points have been plotted the curve should be drawn as smoothly as possible, as shown in Figure 6.15:

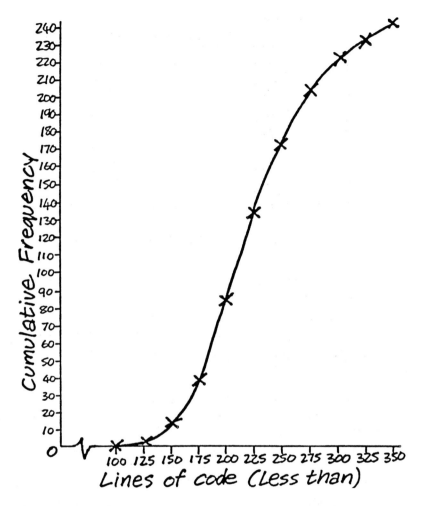

Figure 6.15 Cumulative Frequency Curve

Note that the horizontal axis has been broken at the beginning to allow for a better choice of scale.

Such a curve can now be used to determine how many programs contain less than any given number of lines of code by reading off the values; other uses for such a curve will be introduced later.

6.8 TIME SERIES

A commonly experienced need is to be able to record the way in which a particular variable changes with the passage of time. For example, you may have done such a task at school in recording daily temperatures or rainfall and observing how they changed over the year. In a more commercial example you may wish to record the sales figures of a product line on a week-by-week basis in order to observe the pattern of sales, where peaks and troughs exist, and any evidence of a pattern which can be related to the time of year or the season, etc.

To represent this graphically a time series or historigram (not to be confused with histogram) is used, with time measured along the horizontal axis and the values of the variable occurring on the vertical axis, as shown in Figure 6.16.

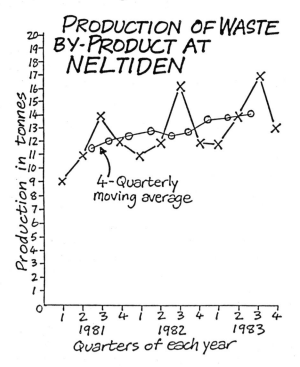

Figure 6.16   Time-series or Historigram

Points are always plotted against the time to which they relate or, if they are average values, then against the mid-point of the time interval or, if they are totals for a period of time, against the end-point of the time interval. Since points other than those plotted have no meaning, straight lines are used to join up the points plotted.

Time series graphs show up seasonal variations, ie situations in which values of the variable tend to peak or to trough at fairly regular intervals, for example, ice-cream sales reaching their highest values in July and August, or the sales of fresh meat being lowest on Mondays. It is possible to suppress these seasonal variations by the use of a suitable moving average, the one shown in Figure 6.16 being a four-quarterly moving average. Note that these are plotted at the centre of the time interval to which they relate and are calculated by *averaging out* the four consecutive figures which cover the year, ie quarter 1 to quarter 4 in 1983, then quarter 2 of 1981 to quarter 1 of 1982, etc. These points are joined together by a smooth curve which shows the trend or the way in which the variable really is changing over the passage of time, without the confusions caused by the peaks and troughs of the seasonal variation.

Other factors, not considered here, can play a part in time series analysis (such as cyclic variation and other special circumstances).

6.9 SCATTER DIAGRAMS

Just as with time series where you wish to show how one variable changes over time, so you may also wish to investigate how one variable changes with respect to a second variable. The kind of distribution which leads to such a situation consists of a set of pairs of values and is called a *bivariate distribution,* with each pair occurring once only in a simple case. Figure 6.17 illustrates such a distribution together with the associated scatter diagram. Note that the two variables are indicated along the two axes and plotted according to the contents of the bivariate distribution.

The scatter diagram itself is no more than a set of points scattered across the graph (hence the name). It is unlikely that these points will lie conveniently along some curve or straight line, but it is certainly likely that they will approximate to one. Generally, if

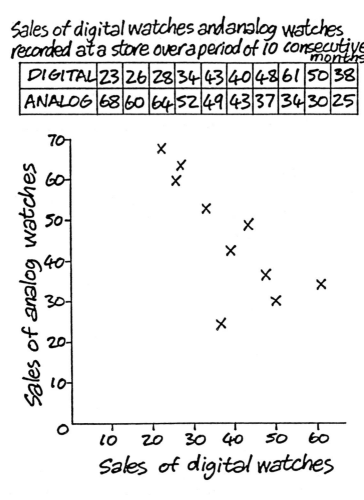

Sales of digital watches and analog watches recorded at a store over a period of 10 consecutive months

DIGITAL	23	26	28	34	43	40	48	61	50	38
ANALOG	68	60	64	52	49	43	37	34	30	25

Figure 6.17   Scatter Diagram

there is evidence that the high values of one variable are associated with the high values of the other variable (and low values with low), then the variables are described as being directly correlated, thus implying that an increase in one will probably cause an increase in the other. If, however, high values of one variable are associated with low values of the other, and vice versa, then the variables are described as being inversely correlated, implying that an increase in one will produce a decrease in the other. Sometimes no obvious pattern emerges at all, in which case there is no correlation between the two variables.

The study of correlation between two or more variables leads on to much other statistics including some very involved calculation and to the associated work on regression beyond the scope of this book.

6.10 EXERCISES

1. An analysis was conducted of the ages of the 120 staff who work in a computer department with the following results:

Age	No of Staff
20–	12
25–	17
30–	21
35–	18
40–	14
45–	13
50–	12
55–	9
60–65	4

Produce a cumulative frequency distribution table from the above data and so construct a cumulative frequency diagram. Use the diagram to find how many employees were aged:

(i) less than 33;

(ii) less than 47;

(iii) between 33 and 46 inclusive.

2. Use the following information about machine downtime to draw up a time series graph and to identify any patterns which may emerge:

Downtime in minutes per day

	Mon	Tues	Wed	Thurs	Fri	Sat
Week 1	97	23	68	68	93	71
Week 2	48	6	86	104	93	51
Week 3	71	45	71	64	87	119
Week 4	115	51	123	118	128	106

3. The following bivariate distribution links the number of times that a system has been run since it was first tested (x) with the number of errors it is known to contain (y) for twelve well-documented systems:

System	A	B	C	D	E	F	G	H	I	J	K	L
x	17	83	62	31	75	91	51	49	12	34	81	9
y	18	6	12	15	8	4	7	11	21	16	5	12

Construct a scatter diagram and comment on any relationship between x and y.

4. The following bivariate distribution shows the relationship between the mean temperature in °C (x) in the machine room and the number of minutes of downtime recorded (y) for each of 10 consecutive days:

Day	A	B	C	D	E	F	G	H	I	J
x	20	21	18	22	17	20	21	23	17	24
y	5	8	10	11	13	4	9	13	14	16

Construct a scatter diagram and comment upon any relationship between the temperatures and the amount of downtime.

7 The Concept of Probability

7.1 AN INTRODUCTION TO PROBABILITY IN DATA PROCESSING

There are several aspects of the work in a data processing department which will, sooner or later, involve the use of probability. For example, if conducting some computer simulations (where the computer is used to create models of situations which are either too dangerous, too time-consuming or too difficult to produce in real life) then to be able to assess the likelihood of each of a number of possible outcomes developing at various stages of the simulation is a requirement. Equally well, in many aspects of operational research there may be a need to be able to calculate probabilities to provide the 'best' information to management engaged in making decisions regarding the viability of some proposals, or calculating how long certain projects are likely to take.

When determining the size and nature of back-up equipment to guard against machine downtime in critical real-time situations such as the computerised control of some industrial process, there is a need to be able to quantify the risks involved and to be able to assess the respective likelihoods of machines (or software) failing. The study of probability theory is, in fact, a very extensive one but only a brief introduction is given in this chapter.

7.2 RELATIVE FREQUENCY HISTOGRAMS

When histograms were introduced in Chapter 6, they were used to allow for comparisons to be made between the frequencies achieved by different categories within a particular distribution. Using these, it was not possible to compare one category in a distribution with a different category in another distribution since the two distributions might be of totally different sizes. To appreciate this more fully consider the following two situations:

(i)   Computer A has 25 hours of downtime out of 2000 hours of running.

(ii) Computer B has 6 hours of downtime out of 100 hours of running.

Clearly it cannot be inferred that B is more reliable than A simply on the basis of their recorded downtime, since the figures do not reflect the total running time in the two cases. In fact, it might well be inferred that A is the more reliable, since it has 1.25% of downtime as against the 6% of machine B. In a more widespread case of trying to compare many different categories in two or more distributions, the use of a *relative frequency histogram* based upon the same idea as the percentage comparison just used, will help to remedy the defects identified in the more straightforward histogram.

In order to create a relative frequency histogram the relative frequency for each category needs to be found; this means that the actual frequency must be divided by the total of all the frequencies for the whole distribution. Hence if one category has a frequency of 12 and the total of all frequencies in the distribution is 200, then that particular category would have a relative frequency of $^{12}/_{200} = 0.06$. Once this has been calculated for each of the categories, the histogram may be drawn using the relative rather than the absolute frequencies. The total of all the relative frequencies for a given distribution will always be exactly 1 no matter what the distribution is, how many categories it may contain, or what the total of the actual frequencies (absolute rather than relative) may have been.

This means that individual parts of two relative frequency histograms are now capable of comparison; for example, those shown in Figure 7.1 which show the ages of equipment in each of two workshops W6 and W8. It can be seen at once that there is a far higher *proportion* of equipment aged between 4 and 5 years old in workshop W8 (a relative frequency of 0.35, or 35% as a percentage) than in workshop W6 (where the relative frequency is 0.1). Had the *actual number* of machines of that age been relied upon then the

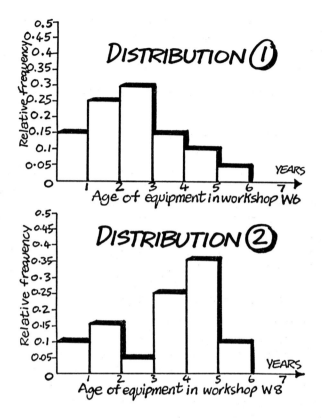

Figure 7.1   Relative Frequency Histograms

respective sizes of the two workshops and the quantity of equipment in each might have affected any conclusions reached.

7.3   THE RELATIONSHIP BETWEEN RELATIVE FREQUENCY AND PROBABILITY

The relative frequency previously introduced measures the way in which one particular category occupies the distribution as a whole and in what proportion. We could as easily ask the question "How likely is it, that a piece of equipment chosen at random from workshop W8, will be between 4 and 5 years old?" and the answer would be the proportion of workshop W8 equipment that is between

4 and 5 years old; such a likelihood is called a *probability* and in this case it has a value of 0.35.

By considering the relative frequency histogram and observing that it is really measuring the probability for each category it becomes evident that the smallest value any probability can take is zero (as in the case of 6 to 7 year old equipment) up to a maximum value of 1 if all the equipment in the workshop were in the same age range.

The phrase "chosen at random" is taken as meaning that all pieces of equipment were equally likely to have been chosen and that there was no reason to suppose that any preference (or bias) would be given to any particular piece or group of pieces. If the equipment were selected by entering the workshop and taking the piece nearest to the door, then it is possible that this might create bias as it is possible that this might be the position usually occupied by the last piece to have been put into the room and, as such, the youngest. Alternatively neither the size nor the colour of the equipment must be allowed to influence the choice.

An ideal way of choosing might be to label each piece with a number and to draw that number at random from corresponding numbers in a box. This offers a reasonably fair way of avoiding bias, if the machine whose number is 'drawn' is identified as the one chosen 'at random' from the workshop.

7.4   ELEMENTARY PROBABILITY AND DEFINITIONS

Probability may formally be defined by reference to the statement given in Figure 7.2.

The idea of an 'experiment' may initially appear to be an odd choice of phrase; it refers simply to any situation in which one of a number of outcomes is possible. Consideration of a few simple examples may clarify the definition.

First, imagine a pack of 52 playing cards from which can be chosen at random any one card which could be any of the four suits, hearts, clubs, diamonds or spades. The experiment refers simply to selecting a card from the pack and the outcomes are the different suits that can be chosen. Given the same experiment one may be more selective and choose to pick a particular face value such as an

If an 'experiment' has N equally likely outcomes and an 'event' is made up of E of these outcomes, we define the probability of that 'event' to be

$$\frac{E}{N}$$

$$\therefore P = \frac{E}{N} \text{ where } p \text{ is the probability of success}$$

Figure 7.2   Elementary Probability Definition

ace, king, queen, etc these being the outcomes in this instance. Since there are 52 cards ($E=52$) of which 13 are spades ($N=13$) the probability of drawing a spade is $^{13}/_{52}$ or $^1/_4$. Similarly, since there are four kings ($N=4$) in the pack the probability of drawing a king is $^4/_{52}$ or $^1/_{13}$. Since there is only one king of spades ($N=1$) the probability of drawing it is $^1/_{52}$.

Taking a different situation, in rolling a traditional six-sided dice, there are six equally likely outcomes. Hence the probability of throwing a 5 is $^1/_6$ (since $E=6$ and $N=1$). If faced with finding the probability of throwing an even number ($N=3$ since there are three such outcomes possible), then the value of that probability would be $^3/_6$ or $^1/_2$. Not surprisingly, the probability of throwing a 7 would be 0 (since $N=0$), a value representing total impossibility. Alternatively finding the probability of throwing a number between 1 and 6 inclusive ($N=6$) would be $^6/_6$ or 1, a value which represents absolute certainty.

The introduction of the word 'success' may convey the attainment of some desired outcome, such as the throwing of an even-numbered face on the dice; thus probability in the manner defined will imply the probability of achieving that success. This probability is denoted by the use of the letter p. Furthermore we denote by the letter q the probability of 'failure', ie of failing to achieve 'success'. Since it is absolutely certain that if an event is not a success then it must be a failure, it follows that the respective probabilities *must* add up to 1, since success and failure are seen as complementary events. Hence:

$$p + q = 1$$

As a clarification of this, if the experiment involved the throwing of the six-sided dice and a success was the throwing of a 3 then (with $N=6$ and $E=1$) $p = 1/6$. A failure must mean failing to throw a 3, ie throwing instead any other number so that (with $N=5$ and $E=6$) $q = 5/6$. This may help to reinforce the point that $p + q = 1$. On the point of notation the expression $P(E)$ is often used to stand for the Probability of achieving an Event, with the particular event often defined in the bracketed expression. Unless there is any possibility of confusion we shall take $P(E) = p$. In those cases where some confusion *is* possible it is better to use subscripted values such as p_1 or p_2 or p_3, etc to refer to different probabilities in the same experiment.

Given that an experiment may be carried out not just once but on a number of different occasions, each is called a 'trial'. Hence, if the dice is thrown on 12 occasions (whatever outcomes they may produce) there have been 12 trials of the same experiment. It may never be declared in advance (with any certainty) what the outcome of any given trial will be, but we can, by theory, by practical work or a mixture of both, identify all the possible outcomes and attach to each one a probability which indicates which outcomes are either more or less likely. If one declares the probability of getting two heads when two coins are thrown is $1/4$, then one can be confident that while the event may not actually happen once on every four throws, it will *tend* to happen on a quarter of all throws as the number of trials gets larger.

7.5   EXERCISES

1. A programmer has noted the results of writing 20 programs. The results indicate that:

      2 compiled correctly on the first run;

      7 compiled correctly on the second run;

      5 compiled correctly on the third run;

      4 compiled correctly on the fourth run;

2 compiled correctly on the fifth run.

On the basis of the foregoing data, calculate the probability that his next program will compile correctly:

(i)   on the first run;

(ii)   on the second run;

(iii)   on the third run;

(iv)   before the fifth run;

(v)   after the third run (but not earlier);

(vi)   at all;

(vii)   but not on the first run.

2.   The operations manager keeps a log of the downtime records of the 30 VDUs in her department. These indicate that:

17 VDUs have never failed;

8 VDUs have each failed on one occasion;

4 VDUs have each failed on two occasions;

1 VDU has failed on three occasions.

On the basis of this data what is the probability that if a VDU were chosen at random from the 30, it would:

(i)   never fail;

(ii)   fail once;

(iii)   fail twice;

(iv)   certainly fail;

(v)   fail twice or three times;

(vi)   fail four times.

7.6 COMBINING PROBABILITIES "BOTH . . . AND . . ."

Events can sometimes be the result of combining two or more events together in some way. The resulting events are then called *compound events*. For example, you may want to know the probability that a system will fail at the *same time* as the person who designed it is away on holiday; these are two quite independent events, but the probability of it happening can be calculated and attributed separately to each one. This type of probability is characterised by the presence of two (or more) events which must happen at the same time. This might be described as "... *both* event A *and also* event B ...". It is assumed that A and B are independent events so the result of A happening in no way alters the probability that B will happen and vice versa. If they are not independent then a slightly different approach to solving the problem has to be used.

However, given that they *are* independent the probability of them both happening is given by the product of their respective probabilities:

$$P(A \text{ and } B) = P(A) * P(B)$$

To illustrate this, it may be useful to return to the earlier example of throwing a six-sided dice. If two trials were conducted then the result of the first can have no effect whatsoever upon that of the second; as such, therefore, these are two independent events. If you were to try to find the probability of throwing a 5 on the first throw and a 2 on the second, then you should multiply the respective probabilities together to give $1/6 \times 1/6 = 1/36$ so that the required probability is $1/36$. Consider next the data processing example suggested earlier. Given the probability that the system will fail is known to be $1/20$ and that that of the system designer being on holiday is $1/10$, the probability that both will happen at the same time is $1/20 \times 1/10 = 1/200$. In the same way the probability that the system will fail when he is *not* on holiday is $1/20 \times 9/10 = 9/200$ since the probability of him not being on holiday must be $1 - 1/10 = 9/10$ (ie 'failure' of the 'he is on holiday' event).

As an extension of the last example, and given the probability that the engineer is out on a call is $2/3$, then the probability of a system failure when both the designer is on holiday and the engineer is out on a call is $1/20 \times 1/10 \times 2/3 = 1/300$, a very small value, so quite an unlikely (but not impossible) event.

Consider the probability that the system will fail and the designer is not on holiday and the engineer is not out on a call. This will be $1/20 \times 9/10 \times 1/3$ or $3/200$.

7.7   EXERCISES ON COMPOUND PROBABILITIES

A microcomputer is being delivered from a local supplier. The probability that they will send the wrong model is $1/5$. There is a probability of $3/4$ that the machine delivered will be working properly. There is a probability of $1/3$ that there will be a fault on the system disk. Calculate the probabilities that:

(i)   the correct model has been delivered;

(ii)   the correct model has been delivered but is not working properly;

(iii)   the correct model has been sent and there is no fault on the system disk;

(iv)   the wrong model has been sent but it is working properly;

(v)   the wrong model has been sent and it is not working properly and there is a fault on the system disk;

(vi)   the correct model has been sent and that it is working correctly and there is no fault on the system disk;

(vii)   the supplier sends the wrong model on two consecutive occasions;

(viii) the supplier sends the wrong model on two consecutive occasions but then sends the correct model on the third delivery.

7.8   COMBINING PROBABILITIES "EITHER . . . OR . . ."

Another type of compound probability is that in which one or other of two events is to occur but not necessarily both. This situation can be recognised by a statement of the kind ". . . *either* event A *or* event B . . ." which will actually include the case of both events occurring at the same time. This can be extended to cases involving more than two events, again leading to the probability that one or more of them takes place (it does not matter how many as long as it is at least one of them). However this time, the events have to be what is described as *mutually exclusive;* for example, when tossing a coin 'heads' and 'tails' are mutually exclusive since the success of one automatically guarantees the failure of the other.

To exemplify this look back at the systems failure situation. The probability that either the designer is on holiday or the engineer is out on call is given by $1/10 + 2/3 = 3/30 + 20/30 = 23/30$. In the same way the probability that the system will fail *or* the designer is on holiday *or* the engineer is not out on a call is given by $1/20 + 1/10 + 1/3 = 3/60 + 6/60 + 20/60 = 29/60$.

Considering the possible causes of the system failure we may identify that Executive will be unable to cope with an interrupt with a probability of $1/15$ (event A) or that a hardware failure may occur at the disk controller (event B) with probability $1/20$. There is furthermore the possibility that the failure is caused by the inability of the system to partition main memory (event C) with a probability of $1/30$. (The actual probabilities quoted here are far higher than might be realistically likely on a real system but their chosen values make for easier calculation and it is convenient to assume here that they are mutually exclusive.) The probability that the system has failed because of either event A or event B is $1/15 + 1/20 = 4/60 + 3/60 = 7/60$. The probability that it has failed because of either event A or event B or event C is given by $1/15 + 1/20 + 1/30 = 4/60 + 3/60 + 2/60 = 9/60 = 3/20$; this figure will include the possibility that two or even all three events have occurred.

7.9 EXERCISES USING COMPOUND PROBABILITIES

1. In a data validation program, fields may be rejected due to one or more of three possible causes; these are, with their respective probabilities of occurrence:

    (a)   wrong data type, $1/6$;

    (b)   wrong value range, $1/5$;

    (c)   wrong format, $1/10$.

    Calculate the probabilities that a field will be rejected because of:

    (i)   wrong data type or wrong value range;

    (ii)   wrong data type or wrong format;

    (iii)   wrong value range or wrong format;

    (iv)   wrong value range or wrong data type or wrong format.

Calculate also the following probabilities:

(v)   the field is not rejected at all;

(vi)   the field is rejected for all three reasons.

2.   Candidates in a data processing examination may either fail, pass, obtain a merit or obtain a distinction with probabilities of 0.2, 0.5, 0.2 and 0.1 respectively.

What are the probabilities of:

(i)   not failing at all;

(ii)   either obtaining a merit or a distinction;

(iii)   failing to obtain either a merit or a distinction?

There are two candidates Subash and Jean both taking the examination.

What are the probabilities that:

(iv)   both fail;

(v)   Subash fails and Jean obtains a pass;

(vi)   Subash fails or Jean gets a distinction;

(vii)   Subash fails and Jean passes or Subash passes and Jean fails;

(viii) Subash does not fail but Jean does or both get merits?

7.10 PROBABILITY TREES

The probability tree provides a diagrammatic representation of a compound probability situation. Consider the two events that a system crashes with a probability of $1/10$ and that the chief operator is away ill with a probability of $3/14$ (shown in Figure 7.3). Drawing up a probability tree shows *all* the possible outcomes of the experiment, including those which depend upon the system actually crashing or otherwise and the operator being away ill or not. There are four possible outcomes from these two events and in the following probability tree these four are seen at the right-hand side with their respective probabilities calculated. In this case only two events are involved but a probability tree can easily be extended to cater for

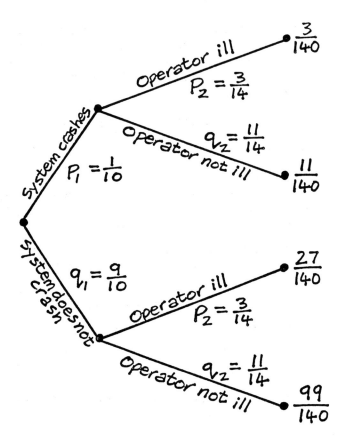

Figure 7.3 Probability Tree

more events but may become unwieldy as the number of events increases. It is quite common when dealing with two events, as here, to use p_1 and p_2 to refer to their 'success' probabilities and to use q_1 and q_2 for the probabilities of failure respectively.

Since it is absolutely certain that one of the four outcomes must be achieved, it is a useful check to see that the four probabilities do add up to 1 as here.

Using this tree the probability that the system will crash at the same time that the chief operator is not away ill can be seen as $^{11}/_{140}$.

You can establish that the probability that either the system crashes when the operator is not ill or the system fails to crash when the operator is ill is given by $^{11}/_{140}$ + $^{27}/_{140}$ (by the addition rule) or $^{38}/_{140}$ which is the same as $^{19}/_{70}$. The order in which they are presented is unimportant.

7.11 EXERCISES INVOLVING PROBABILITY TREES

1. Produce a probability tree to show the four outcomes and their respective probabilities involving two events. The first event is that a microcomputer has been delivered from a local supplier but that they have sent the wrong model with a probability of $^1/_5$. The second event is that the machine delivered will be working properly with a probability of $^3/_4$.

2. Extend the tree diagram in question 1 to include the third event that the system disk will be faulty with a probability of $^1/_3$.

3. Produce a probability tree to show the eight outcomes of the experiment including the three events that:

    a system fails with a probability of $^1/_{20}$;

    the system designer is on holiday with a probability of $^1/_{10}$;

    the engineer is out on a call with a probability of $^2/_3$.

(Check that the probabilities of these eight outcomes as shown on the tree do indeed total 1.)

8 Measures of Average

8.1 INTRODUCTION

The word 'average' is, to say the least, a very loosely defined term. It refers, in general, to the taking of a single value or quantity, to be representative of, usually, a large number of such values or quantities; as such it is frequently referred to as a 'measure of central tendency'. The problem is that there is no single method of obtaining such a value, so that a choice has to be made from a number available. The choice will be determined partly by the nature of the values to be represented, be they qualitative or quantitative etc and partly by the purpose for which the average is required.

In elementary mathematics courses the word 'average' is usually synonymous with what we shall define as the arithmetic mean; however we shall see that other alternative averages exist, such as the median and the mode. Furthermore the often-used approach of referring to the arithmetic mean simply as the 'mean' is not sufficiently explicit, since there is also the geometric mean and the harmonic mean, which both provide alternative methods for measuring an average. These extra two are not dealt with in this chapter however, which is restricted to arithmetic mean, median and mode. Any further reference to 'mean' in this book will refer to arithmetic mean only to avoid any likelihood of confusion.

8.2 THE MEAN OF A SET OF DATA ITEMS

If presented with the eight numbers:

$$7 \quad 21 \quad 13 \quad 17 \quad 23 \quad 18 \quad 9 \quad 20$$

and added them up (giving 128), then divided the total by the

number of items (8) the value of the mean obtained would be
$^{128}/_8 = 16$. This particular average can clearly only be used with data
which is quantitative. In working it out, every one of the original
eight values was taken into account so that each one will have had an
equal influence on the value obtained for the mean. It is worth
mentioning here that the mean may not be equal to any of the
original numbers at all (as in this case), but may do so on other
occasions. Note however that the presence of even one value which
is disproportionate to the others in terms of its size may easily
distort the value of the mean obtained. For example if the original
set of values was the 12 numbers:

$$3 \quad 3 \quad 4 \quad 4 \quad 4 \quad 4 \quad 5 \quad 5 \quad 6 \quad 6 \quad 6 \quad 176$$

then the mean would be the total (226) divided by the number of
items (12) giving 18.83 (to two decimal places). This value would
not have been seen as being in any way typical of the first 11 items in
the set, as the twelfth item, 176, was so much out of keeping with all
the others as to distort the picture. This 'liability to distort' is one of
the factors to bear in mind when attempting to decide which average
is to be used. An obvious application is to look at the wages paid to
25 staff working for a certain business. Of these one is paid $500 a
week, 8 get $150 a week and the other 16 are paid $100 a week.
Establishing an average in this way by adding together the three
values $500, $150 and $100 (giving $750) and dividing by 3 gives a
'mean' of $250 a week. This figure, supposedly representative of the
25 wages, is in fact larger than 24 of them, caused by the unbalancing
effect of the solitary $500 value. However, it is possible to
compensate for this by 'weighting' the wages in accordance with the
number of people receiving each value, so that the $500 would get a
weighting of one and the value of $100 would get a weighting of 16;
this has the effect of reducing the impact of the single $500 and in so
doing produces a far more realistic value for the mean. This 'weighted'
aspect is further discussed later on in this chapter.

In the diagram given in Figure 8.1 a computer-relevant example is
given in which the average speed of key-punch operators in the data
preparation section is required. Since a formula needs to be
established for the calculation of the mean for future use, the letter
x denotes a typical speed so that there are 14 different x-values in
this example. The notation Σx (pronounced sigma x) denotes the

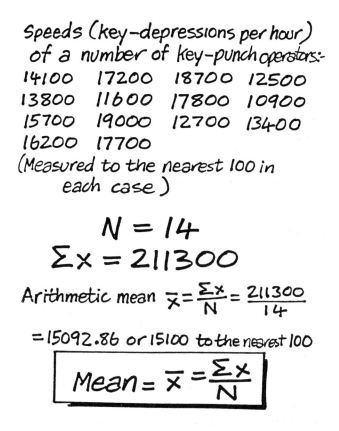

Speeds (key-depressions per hour)
of a number of key-punch operators:-

14100	17200	18700	12500
13800	11600	17800	10900
15700	19000	12700	13400
16200	17700		

(Measured to the nearest 100 in each case)

$$N = 14$$
$$\Sigma x = 211300$$

Arithmetic mean $\bar{x} = \dfrac{\Sigma x}{N} = \dfrac{211300}{14}$

$= 15092.86$ or 15100 to the nearest 100

$$\boxed{\text{Mean} = \bar{x} = \dfrac{\Sigma x}{N}}$$

Figure 8.1 Establishing the Mean of a Set of Data Items

result of adding up all the x-values ie the total of the x-values. Additionally N is used to denote the number of x-values there are (14 in this case). Finally the notation $\bar{x}$ refers to the mean of the x-values (read as x-bar).

In this example note that the original values were quoted 'to the nearest 100' and for this reason the $\bar{x}$-value has been rounded so that it too is given to the same level of accuracy; it makes no sense at all to quote a calculated mean to any level of accuracy greater than that of the original data. The mean has exactly the same units as those of the original data and it is worth noting that, in order to calculate a mean, the data must first be quoted in the same units as one another; ie do not mix minutes with seconds or hours unless they are converted to the same common set of units.

Having obtained the figure of 15,100 key-depressions per hour you might use it to estimate the time it might take six of these operators, chosen at random, to punch a volume of data of 500,000 characters:

Average output of 6 operators = 6 × 15,100 = 90,600 characters per hour

$500,000/90,600$ = 5.52 hours or about $5^{1}/_{2}$ hours.

Following on from the same set of circumstances there may be a second group of 29 key-punch operators whose average speed is only 12,300 key-depressions per hour and you want to find the average speed of the combined group (29 + 14 = 43 operators). Although the individual speeds of the 29 are not known, it can be calculated that the total of their 29 speeds must be 29 × 12,300 = 356,700 (by reversing the calculation of $\bar{x}$). The total of the speeds of the first 14 operators is already known to be 211,300; hence the total of the speeds of all 43 operators must be 356,700 + 211,300 = 568,000 so the average must be $568,000/43$ = 13,209 or 13,200 key-depressions per hour when rounded to the nearest 100. Note that the value is *not* the result of adding together the two means of 15,100 and 12,300 and dividing the result by two, since it must take into account the number of people in each group. Hence the result of 13,200 is rather closer to 12,300 than to the 15,100 since there are more people in the slower group.

8.3   EXERCISES ON MEANS

1.   A check was made to find how many boxes of listing paper had been used each day during a seven-day period. The figures recorded were:

12   14   17   16   24   13   7

(a)   Calculate the average number of boxes used per day.

(b)   Calculate the number of days that a delivery of 400 boxes is likely to last.

2. The response times of 11 VDUs linked up to a mainframe computer were each recorded to the nearest 0.1 second as follows:

   2.3 1.6 2.9 3.6 4.1 2.6 2.7 3.1 3.8 2.7 2.8

   (a) Calculate the mean response time.

   As the result of an amendment to the system software the following response times were recorded for the same 11 VDUs:

   1.8 1.3 3.1 3.2 3.8 2.8 2.7 3.0 3.5 2.1 1.8

   (b) Calculate the new mean response time.

   (c) Using your results for (a) and (b) calculate the percentage reduction in response time caused by the changes made to the system software.

3. One group of 12 programmers with a mean experience of 5 years of Cobol were merged with a larger group of 17 who had a mean experience of 7.2 years. Calculate the mean experience level of the combined group.

8.4 THE MEDIAN OF A SET OF DATA ITEMS

Given the same data about key-punch operators as used in the section on the mean, it might be asked that you select one person as a 'typical' member of the section to test some new equipment for its ease of use. Clearly it is not advisable to go for the fastest nor for the slowest but rather to select somebody whose speed is somewhere 'near the middle'. A quick examination of the data will soon show that not one of the 14 staff has a speed exactly equal to the calculated mean (and such a situation, as already noted, is likely to be the most usual) so a different approach is called for as follows:

> Arrange the values in order of magnitude, lowest to highest, and select as the average the one that occurs in the middle of the list, calling it this time the median.

If there are N items the median will therefore be the $\frac{1}{2}(N+1)$th value. Should N be an odd number this will identify one value uniquely, but if N is even, as in the key-punch operator case ($N=14$), then the median is the $7\frac{1}{2}$th value, which is interpreted as being exactly half-way between the 7th and the 8th values. The 14 values, arranged in order of magnitude, are:

10900 11600 12500 12700 13400 13800 14100 15700 16200

17200 17700 17800 18700 19000

The 7th and 8th values are, respectively, 14100 and 15700 so the median is $1/2(14100 + 15700) = 14900$.

The illustration shown in Figure 8.2 takes a case relating to processor sizes of microcomputer in which the value of N is odd:

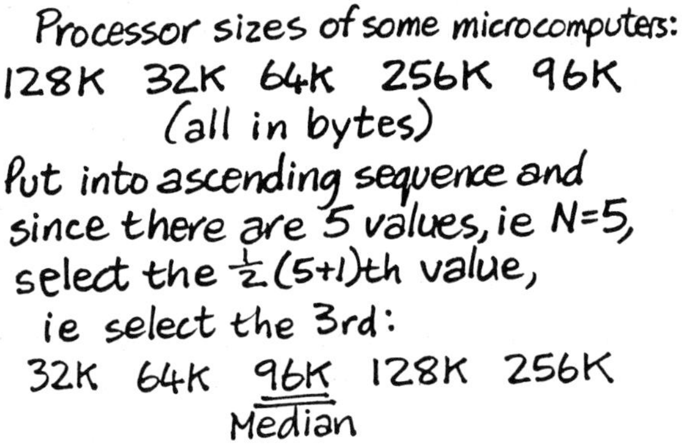

Processor sizes of some microcomputers:
128K 32K 64K 256K 96K
(all in bytes)
Put into ascending sequence and
since there are 5 values, ie N=5,
select the $\frac{1}{2}(5+1)$th value,
ie select the 3rd:
32K 64K 96K 128K 256K
Median

Figure 8.2   Example of Median of a Set of Data Items

Benchmark test times are the subject of the case in Figure 8.3, in which the value of N is even.

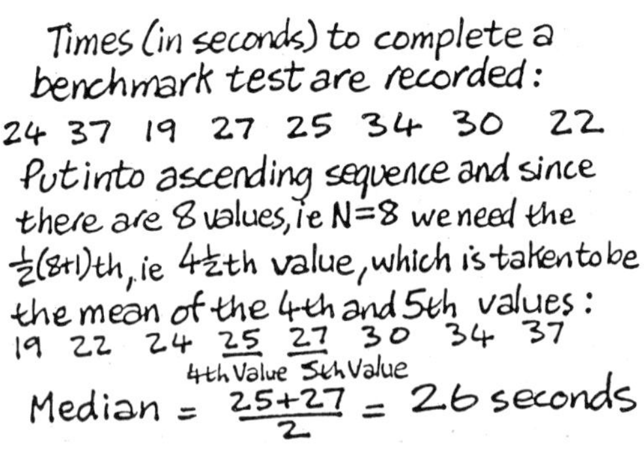

Times (in seconds) to complete a
benchmark test are recorded:
24 37 19 27 25 34 30 22
Put into ascending sequence and since
there are 8 values, ie N=8 we need the
$\frac{1}{2}(8+1)$th, ie $4\frac{1}{2}$th value, which is taken to be
the mean of the 4th and 5th values:
19 22 24 25 27 30 34 37
      4th Value 5th Value
Median = $\frac{25+27}{2}$ = 26 seconds

Figure 8.3   Example of Working out the Median of a Set of Data Items

The median does possess some particular advantages; it is more easily found than the mean and it is not so affected by extreme values. The median of the seven items:

3   4   7   8   12   15   8164

is exactly the same as the median of the seven items:

3   4   7   8   12   15   17

This is a disadvantage however at the same time, since it means that the value of the median does not at all reflect the values of anything but the 'middle' term(s) in a data set, whereas the mean is affected by each value.

If it had not been possible to quantify the data but merely to list it in order of priority or preference, it is still easy to determine the median even if it is impossible to calculate the mean.

8.5   EXERCISES ON MEDIAN

1.   The following are the ages in years of the 13 staff in the systems section of a local computer department:

37   26   29   42   28   51   23   32   49   29   35   41   38

Arrange the ages in ascending order of magnitude and hence calculate the median age.

2.   The run-times (in minutes) of 12 jobs were noted and listed below:

43   28   37   21   16   59   18   24   37   41   12   29

Calculate the median run-time.

8.6   THE MODE OF A SET OF DATA ITEMS

The mode is defined to be that value which appears most commonly in the list of data items, ie the 'most popular' item. As this completely ignores the values of the data items it is especially well-suited to data which is qualitative. In the table shown in Figure 8.4 the popularity of a number of different programming languages was investigated.

If two values appear with equal frequencies, both proving to be more commonly found than any other value, both are classified as

Results of a survey into the number
of companies using different
programming languages

Language used	Frequency
ALGOL	4
BASIC	13
COBOL	27
DIBOL	3
FORTRAN	10
PASCAL	8
RPG II	14
SNOBOL	1

The mode is COBOL since it
has the highest frequency.
Clearly if the survey was conducted with a
different group of companies the results might
be different and the mode might change

Figure 8.4   The Mode of a Set of Data Items

the modes and the distribution is referred to as bimodal.

Apart from its ease of determination, the mode may be used for planning purposes. If (all other factors being equal), a decision had to be made about which language to select for the purpose of training a group of new recruits to computing at some training school or other, then the most popular language (ie Cobol in the previous case) is likely to be the one to be used. If it were possible and desirable to train them in two languages then the most popular, Cobol, and the second most popular, RPG11 would seem to provide the best coverage since the trainees would be capable of working in 41 of the 80 companies in the survey. Note however, that the choice of language to be used may be determined by factors other than straightforward popularity.

8.7 EXERCISES INVOLVING THE MODE

1. Determine the mode of the following set of values:

6	8	5	12	6	8	16	5	9	3	12
4	9	6	5	13	7	9	8	2	5	1

2. When the óperators were asked to choose the colour of the walls in their newly-decorated rest room, the 47 of them chose as follows:

Beige	12
Blue	3
Lemon	8
Mustard	4
Pink	9
White	11

    What was the mode?

8.8 THE MEAN OF A FREQUENCY DISTRIBUTION

In an earlier section the mean was defined in the case of a straight-forward set of data values. The definition is still the same even in the case of a frequency distribution but with some elements of modification. For example, if the x-value 17 occurs with a frequency of 3 (ie $f = 3$) then *each* of the three 17s must contribute to the total. This time therefore the calculation of the mean is presented in a tabulated manner, the columns being headed x, f and fx respectively; fx is the product of the x-value with its frequency f. Thus if $x = 17$ and $f = 3$ then $fx = 51$ and this is the contribution of the value 17 to the grand total. The N used earlier is replaced now by Σf to total up *all* the items in the calculation. The example given in Figure 8.5 shows a table in which the first two columns define the distribution itself and the third column contains the calculated fx values. The revised formula for $\bar{x}$ is now $\Sigma fx / \Sigma f$ in which Σfx replaces the previous Σx and Σf replaces N. One might regard this as a shorthand way of writing out each of the 100 values (hence $N = 100$) and adding them up, which would give a total of 2077 (hence $\Sigma x = 2077$) if all 100 values were identified separately.

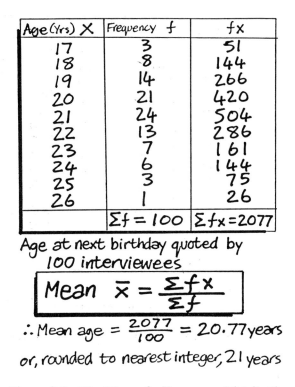

Age (Yrs) X	Frequency f	fx
17	3	51
18	8	144
19	14	266
20	21	420
21	24	504
22	13	286
23	7	161
24	6	144
25	3	75
26	1	26
	$\Sigma f = 100$	$\Sigma fx = 2077$

Age at next birthday quoted by 100 interviewees

$$\text{Mean } \bar{x} = \frac{\Sigma fx}{\Sigma f}$$

$$\therefore \text{Mean age} = \frac{2077}{100} = 20.77 \text{ years}$$

or, rounded to nearest integer, 21 years

Figure 8.5   The Mean of a Frequency Distribution

There is a useful device that can be used to reduce still further any tedious arithmetic, especially in cases where the x-values are large and in which the fx calculations would start to become unwieldy. In this case decide upon a 'working zero' or 'assumed mean' which is denoted by x_0 and is chosen to be somewhere near where you may expect the mean to actually lie. The accuracy of the choice is relatively unimportant; what is far more important is that it is chosen so that when x_0 is subtracted from each of the x-values, the resulting $x - x_0$ column or d column (using d for the difference $x - x_0$) contains as simple a set of values as possible (ideally integers), some positive, some negative, so that the final fd column calculation is simplified. This time fd is used instead of the fx column in the previous calculation (Figure 8.6). Since every x-value has been reduced by x_0 the formula for $\bar{x}$ adds x_0 back on again afterwards.

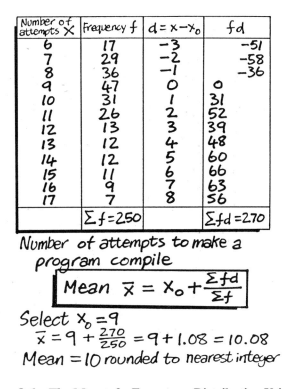

Number of attempts X	Frequency f	$d = x - x_0$	fd
6	17	−3	−51
7	29	−2	−58
8	36	−1	−36
9	47	0	0
10	31	1	31
11	26	2	52
12	13	3	39
13	12	4	48
14	12	5	60
15	11	6	66
16	9	7	63
17	7	8	56
	$\Sigma f = 250$		$\Sigma fd = 270$

Number of attempts to make a program compile

$$\text{Mean } \bar{x} = x_0 + \frac{\Sigma fd}{\Sigma f}$$

Select $x_0 = 9$

$\bar{x} = 9 + \frac{270}{250} = 9 + 1.08 = 10.08$

Mean $= 10$ rounded to nearest integer

**Figure 8.6   The Mean of a Frequency Distribution Using a
Working Zero or Assumed Mean**

Note that, since the *fd* column is likely to include both positive and negative products having them separately columnised reduces the risk of miscalculation. In Figure 8.6 Σfd equals the sum of the positive *fd* values less the negative *fd* contributions ($= 415 - 145 = 270$). Frequently the degree of calculation is minimised by choosing to use as x_0 the value of *x* with the highest frequency (ie the modal value) and that was done here; whilst it does not *always* lead to the simplest arithmetic it is a fairly reliable choice.

There is one further step that may be taken, usually in cases when dealing with grouped frequency distributions in which the class widths are equal. In such cases the mid-interval value of each class is taken as the *x*-value, the rest of the calculations following as before.

When the *d* column is obtained, it is likely that it will consist of a set of values with a common factor (usually the same as the class width). Denoting this common factor or 'unit' by the letter u, the *d*-values can be divided by u (to scale *down* all the numbers in later columns) so obtaining a d'-column (with $d' = d/u$). This time the final column contains fd' values; these are all equally scaled-down versions of what would normally have been fd values, thus simplifying the arithmetic. At the calculation stage for $\bar{x}$ the scaling down is reversed (hence the multiplication by u) *as well as* the adding back on of x_0 in the formula.

In Figure 8.7 u was selected as 5, the most obvious common factor for the *d*-column, after already selecting $x_0 = 37\frac{1}{2}$ on the basis of this being the modal value.

Age range (years)	Mid-interval value (x)	Frequency f	$d = x - x_0$	$d' = \dfrac{d}{u}$	fd'
15–	$17\frac{1}{2}$	12	−20	−4	−48
20–	$22\frac{1}{2}$	14	−15	−3	−42
25–	$27\frac{1}{2}$	23	−10	−2	−46
30–	$32\frac{1}{2}$	31	−5	−1	−31
35–	$37\frac{1}{2}$	33	0	0	0
40–	$42\frac{1}{2}$	28	5	1	28
45–	$47\frac{1}{2}$	22	10	2	44
50–	$52\frac{1}{2}$	15	15	3	45
55–	$57\frac{1}{2}$	12	20	4	48
60–	$62\frac{1}{2}$	8	25	5	40
65–70	$67\frac{1}{2}$	2	30	6	12
		$\Sigma f = 200$			$\Sigma fd' = 50$

Ages of people interviewed in the street

$$\text{Mean } \bar{x} = x_0 + u\frac{\Sigma fd'}{\Sigma f}$$

Select $x_0 = 37\frac{1}{2}$ and $u = 5$

$$\therefore \bar{x} = 37\frac{1}{2} + 5 \cdot \frac{50}{200}$$
$$= 37.5 + 1.25 = 38.75$$

Mean = 38.75 years

Figure 8.7   The Mean of a Grouped Frequency Distribution

In dealing with grouped frequency distributions, particular care must be taken to distinguish between discrete and continuous data when determining mid-interval values.

8.9 EXERCISES INVOLVING THE MEAN OF A FREQUENCY DISTRIBUTION

1. The following table records the number of telephone calls from external users per day (x) at a Help Desk facility during its first 50 days of operation:

No of calls/day, x	No of days, f
17	4
18	8
19	19
20	12
21	5
22	2

Calculate the mean number of telephone calls per day over this period.

2. The table records the number of pages of printed output (x) for each of 100 jobs run at the Operations Centre:

No of pages output, x	No of jobs, f
73	7
74	12
75	15
76	23
77	17
78	12
79	9
80	3
81	2

By choice of a suitable assumed mean, calculate the mean number of pages of output per job.

3. The following table records the number of transcription errors detected each hour, x, for each of 80 recorded hours in the data preparation department:

No of errors, x	No of times found, f
1 – 50	6
51 – 100	12
101 – 150	19
151 – 200	21
201 – 250	16
251 – 300	6

By choice of a suitable assumed mean and unit size calculate the mean number of errors detected per hour.

4. The following table records the ranges of annual salaries of the 200 employees of a large computer bureau:

Salary ($)	Number of employees
3000 –	6
4000 –	7
5000 –	25
6000 –	49
7000 –	56
8000 –	31
9000 –	14
10000 – 11000	12

Calculate the mean annual salary.

8.10 THE MEDIAN OF A FREQUENCY DISTRIBUTION

In all cases in which the data is not grouped it is possible to find the median by inspection as was defined earlier by quoting the $\frac{1}{2}(N+1)$th value when the data is arranged in order of magnitude. For example, if presented with:

x	f
17	3
18	5
19	18
20	36
21	15
22	4

then if the 81 values were arranged as 17 17 17 18 18 18 18 18 19, etc, it is evident that the $\frac{1}{2}(81+1)$th (ie the 41st) will be 20, so that the median will be 20. It so happens that this time the median is equal to the mode; this may happen in a number of distributions but it is *not* generally true.

If, however, the data is grouped so that the actual values of the individual items are not known, an estimate has to be made of the median by producing a cumulative frequency distribution and drawing a cumulative frequency diagram as illustrated in Figure 8.8.

In this case there had been 300 data items so that the median would have to be the $\frac{1}{2}(301)$th value or $150\frac{1}{2}$th value; since N is large the $\frac{1}{2}(N)$th value is generally used instead, thus quoting the 150th value (the discrepancy is too small to bother you unduly).

To obtain the median simply search for 150 on the cf-axis and go across to the curve horizontally, then read down vertically onto the x-axis to read off the median directly, in this case as 156 kg. Two other values that will be of importance in the next chapter may also be introduced here, the lower and the upper quartiles. The lower quartile is the $\frac{1}{4}N$th value (in this case, the 75th) and the upper quartile is the $\frac{3}{4}N$th value (here the 225th) so that the two quartiles together with the median divide up the whole distribution into four equally sized groups. The lower quartile is therefore such that one quarter of all data values are below it (and three quarters above) whilst the upper quartile has three quarters of all data values below

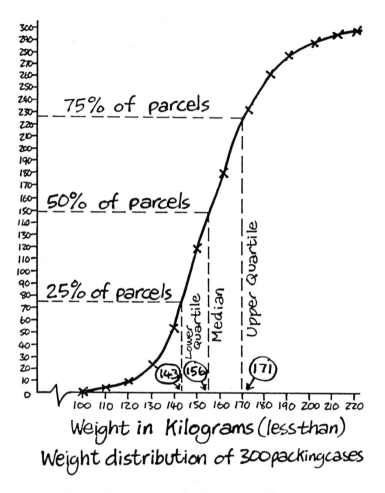

Figure 8.8   Cumulative Frequency Diagram

it (and only one quarter above). In the previous illustration the lower and upper quartiles can be read off as 143 kg and 171 kg respectively.

8.11 EXERCISES INVOLVING THE MEDIAN OF A FREQUENCY DISTRIBUTION

1.   Find the median of the following distribution of estimates made

by 115 operators as to the ideal size for an operations shift team:

Size, x	No of estimates, f
3	16
4	24
5	27
6	29
7	16
8	3

2. By first creating a cumulative frequency table and then a cumulative frequency diagram, estimate the median of the following survey of the salaries of 200 people in a computer installation:

Salary ($)	No of people
4000 –	7
5000 –	19
6000 –	38
7000 –	45
8000 –	34
9000 –	27
10000 –	21
11000 – 12000	9

8.12 THE MODE OF A FREQUENCY DISTRIBUTION

If dealing with data which is not grouped, then the mode, like the median, can be found by inspection; it is quite simply the x-value with the highest value of f. If, however, it *is* grouped then the only conclusion which can be arrived at with any certainty is that the *modal class* (rather than the mode) is the group or category with the highest frequency; since the individual data values within each class are not known it cannot possibly be determined with any justification which one data value occurs more frequently than any other so the mode cannot be found.

To illustrate this point consider the following portion of a grouped distribution involving integer values of x only:

x	f
51–55	12
56–60	16
61–65	10

On the basis of this 56–60 must be the modal class but since the

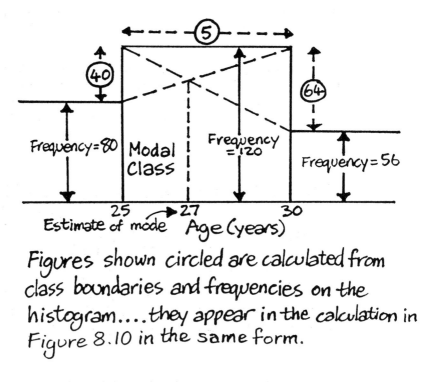

Figures shown circled are calculated from class boundaries and frequencies on the histogram....they appear in the calculation in Figure 8.10 in the same form.

Figure 8.9   Estimating the Mode Using a Histogram

original x-values prior to grouping are not known, where the mode lies cannot be determined. It is quite possible that no one value in the 56–60 grouping occurs more than four times yet it is possible that 63 occurs 10 times so that 63 might be a contender for the mode even though it does not lie within the modal group.

Nevertheless, if asked to make an *estimate* of the mode with no knowledge other than the details of the grouped distribution it would have to be assumed that it did lie within the modal class and that it occupied a position in that class determined by the values in the distribution table. This can be done either graphically or by calculation. The graphical method is shown in Figure 8.9, as a histogram. Note the way in which lines are drawn from the tops of the two cells adjoining the modal class so they intersect vertically above what will now be the estimate of the mode.

The calculation method shown in Figure 8.10 depends upon using the values contained in the histogram, the 25 being the lowest x-value in the 25–30 modal class.

This approach would appear to give a more accurate result; in fact such accuracy is totally spurious since the calculation uses exactly the same quantities and method as in the graphical approach. The figure to be quoted, 27 years, is at best only an estimate of the mode.

$$\text{Estimate of mode} = 25 + \boxed{5} \times \frac{\boxed{40}}{\boxed{40} + \boxed{64}}$$

$$= 25 + 5 \times \frac{40}{104} = 25 + \frac{200}{104}$$

$$= 25 + 1.9 = 26.9 \text{ years}$$

The calculated value would appear to be more 'accurate' than that obtained graphically; since both are only estimates, this difference is not important.

Figure 8.10   Estimating the Mode Through Calculation

8.13 EXERCISES INVOLVING THE MODE OF A FREQUENCY DISTRIBUTION

1. Find the mode of the following distribution:

x	f
92	7
93	19
94	26
95	34
96	15

2. Estimate the mode of the following distribution of ages of staff in the computer department:

Age (years)	No of staff
20–	7
25–	18
30–	21
35–	28
40–	19
45–50	7

(i) By graphical means.

(ii) By calculation means.

8.14 SUMMARY

In comparing the three forms of average, each clearly has some advantages coupled with some disadvantages. The mode is the only one that can be used with qualitative data for example. The mean uses every data value but can easily be distorted by extraneous values; the median and the mode effectively ignore all other data values so are not distorted by them. This also means that whereas the mean is sensitive to changes in the data, however small, the other two are quite insensitive to such changes.

There is a 'formula' which connects the three values but it should at best be regarded as a guideline rather than an accurate relationship. It is:

$$\text{mode} = \text{mean} - 3\,(\text{mean} - \text{median})$$

From a mathematical point of view the mean is the only one which has a 'proper' formula. Not surprisingly therefore it will appear again in the next chapter.

9 Measures of Spread or Dispersion

9.1 INTRODUCTION

The average was introduced as a method of identifying one item to represent or be typical of a group of items; this did indeed meet a well-defined need. However, even having such a solution fails to tell us all that we need to know about the distribution itself because it is quite possible for two sets of values to have exactly the same mean but to be very different in nature. Consider the four distributions shown in Figure 9.1 in which the mean has been calculated for each.

Distribution A: 5, 6, 7, 8, 9, 10, 11
Mean is 8

Distribution B: 5, 5, 5, 8, 11, 11, 11
Mean is 8

Distribution C: 1, 2, 3, 8, 13, 14, 15
Mean is 8

Distribution D: 14, 15, 17, 19, 20
Mean is 17

Figure 9.1   The Mean of Different Distributions

143

The distributions A, B and C each have seven values and the same mean of eight yet they are different in other ways; A involves an even distribution of values between the two end-points of 5 and 11 and although B uses the same end-points the values are spread in such a way as to be heavily bunched at the end-points. In C the end-points are further apart yet the values are still clustered close by them without affecting the mean at all. With D the difference between the end-points is the same as in cases A and B yet the mean is quite different.

A second parameter must therefore be introduced to assist in describing the distribution. This is one which should identify the way in which the values are spread out or dispersed; it must reflect the fact that the values in C are far more *widely* scattered than in the other three cases and it should also distinguish between the uniformity of distribution seen in A and the clustering around the end-points of B. Just as in the case of averages there is no single such measure of dispersion. This chapter will examine a few of the different ways of representing it.

9.2   THE RANGE

The range is by far the simplest measure of dispersion. It is found for any distribution by subtracting the smallest data value in the distribution from the largest. In respect of the four distributions A, B, C and D, the ranges are listed in Figure 9.2.

Distribution A   Range is 6
      "   B   "   " 6
      "   C   "   " 14
      "   D   "   " 6

Figure 9.2   The Range of a Distribution

Immediately we can see that B and C are now more readily distinguished from one another; both have the same mean of 8 but the greater degree of spread of C leads to a range of 14 rather than to one of 6 as in the case of B. It is worth noting at the same time that A and D have the same range yet different means. However, the range totally fails to distinguish between A and B which have equal means and equal ranges but, as already indicated, the data in B is more clustered around the two end-points than with A. Thus the range is incapable of coping with what happens between the end-points. Nevertheless, it provides us with a very easily obtained measure of dispersion even if it is not capable of providing the degree of discrimination required in all cases.

Just as the mean could be distorted by extraneous 'untypical' data items, so too can the range. In the case of distribution A earlier, if one were to replace the value 11 by a value of 17 so that the distribution now becomes:

$$5 \quad 6 \quad 7 \quad 8 \quad 9 \quad 10 \quad 17$$

then the mean would have changed from 8 to $8^6/7$ and the range would have been doubled from 6 to 12. Had the values *between* 5 and 11 of the original distribution A been altered the range would not have changed at all in the process.

9.3 INTERQUARTILE RANGE

In the previous chapter the upper and lower quartiles were introduced when obtaining the median of a grouped frequency distribution. The difference between these values is referred to as the *interquartile range*. Thus in the example quoted when the lower and upper quartiles were found to be 143 kg and 171 kg respectively, the interquartile range is $171 - 143 = 28$ kg, noting that, like the range, it has the same units as the data items.

The interquartile range is a slightly better measure of spread than the range because any 'untypical' values will almost certainly be found in the lowest or the highest 25% of the data values and as such will not distort the calculated value of the interquartile range. However, it still suffers from a large number of the same problems experienced with the range and since it is generally inconvenient to draw up cumulative frequency diagrams *solely* to find the inter-quartile range, it is not very widely used.

If given a distribution which was *not* grouped, then finding the two quartiles is fairly quickly established. They are defined as the

$1/4(N+1)$th and the $3/4(N+1)$th values to give lower and upper quartiles respectively (note that the $1/4N$ and $3/4N$ values quoted in the previous chapter are suitable approximations when the number of data items, N, is large). Hence if 15 items were:

10   3   12   16   5   2   8   17   11   13   6   8   12   16   4

and then they were rearranged in ascending order:

2   3   4   5   6   8   8   10   11   12   12   13   16   16   17

with the lower and upper quartiles underlined (with $N = 15$, lower quartile $= 1/4(16)$th or 4th value whilst the upper quartile $= 3/4(16)$th or 12th value). Hence for this distribution the interquartile range is $13 - 5 = 8$. Should any values below 5 or above 13 be changed, eg the 17 becoming 2139, there will be *no* change in the interquartile range even if the range changes from 15 to 2137 in the process.

9.4   EXERCISES INVOLVING RANGE AND INTERQUARTILE RANGE

1.   Find both the range and the interquartile range of the data values:

29   7   15   23   11   14   36   8   14   17   21

2.   If the value of 36 in question 1 were changed to 39 what effect would it have on:

(i)   the range;

(ii)   the interquartile range?

3.   The semi-interquartile range is defined to be half of the inter-quartile range. Find the semi-interquartile range of the data values:

741   865   637   801   782   871   915

4.   Find for the set of data values:

23   17   29   36   15   6   18   24   32   41   14   27   33   18   24

(i)   the range;

(ii)   the interquartile range;

(iii)   the semi-interquartile range.

5.   If the value of 17 in question 4 were changed to 15 what effect
     would it have on:

     (i)   the range;

     (ii)   the interquartile range;

     (iii) the semi-interquartile range?

9.5   STANDARD DEVIATION AND VARIANCE

A measure of dispersion which can overcome the objections listed
for both the range and the interquartile range is the standard
deviation. At first it may seem that the manner of its calculation is
long and unwieldy, but since it can be calculated almost as a
by-product of the calculation for the mean of a grouped frequency
distribution, in most cases there will be relatively little extra work to
be done.

The points that are required of a measure of dispersion for it to be
viewed favourably are:

(i)   that every data value is taken into account;

(ii)   that data values are weighted according to their frequency.

The standard deviation satisfies these criteria and is calculated
according to the following steps:

(a)   Calculate the mean, $\bar{x}$ of the data items.

(b)   Obtain the deviations $d = x - \bar{x}$ for each of the data items x.

(c)   Obtain the squares of these deviations, d^2, for each value x.

(d)   Find the mean of the d^2 values.

(e)   Obtain the square root of this value.

The result of (e) is the standard deviation of the original set of
data items. The example in Figure 9.3 shows the calculation of the
standard deviation for a distribution of 10 data values.

Note in particular that at stage 3 the deviations have been
obtained; some of these values are positive, some negative, and the
squaring which takes place next is a means of eliminating the signs;
it is compensated for by the last stage in which a square root is
taken.

1. Data values are:
   3 6 7 9 9 11 12 13 14 19
2. Their mean is calculated to be: 10.3
3. Deviations of the data values from
   the mean are: -7.3 -4.3 -3.3 -1.3
   -1.3 0.7 1.7 2.7 3.7 and 8.7 respectively
4. The squared deviations are:
   53.29 18.49 10.89 1.69 1.69 0.49
   2.89 7.29 13.69 and 75.69 respectively
5. The mean of these squared deviations
   is 186.1 ÷ 10   ie 18.61
6. Standard deviation is √18.61 = 4.314

Figure 9.3 Calculation of Standard Deviation

Occasionally reference may be made to the variance; this is simply the square of the standard deviation. In the previous example, the variance was calculated at stage 5 to be 18.61, so it is usually calculated in the very process of calculating the standard deviation rather than as an additional task.

The standard deviation, in common with both range and inter-quartile range has exactly the same units as did the original data values. The variance, on the other hand, has units which are the square of the original units; thus if the data items were all lengths in centimetres (cm) then the variance would automatically be in cm^2.

The standard deviation, together with the mean, provides us with a method of comparing two or more distributions in a workable manner. Together with the concept of probability these two measures form the cornerstone of much statistical work, though not covered in this book. A second example, (in which we are required to find the standard deviation and the variance) is provided by the following set of 11 response times, measured in seconds, found at a number of VDUs linked to a mainframe computer:

5.3   3.2   2.6   2.9   3.4   4.7   6.8   2.9   3.7   5.1   4.8

The total of these 11 values is 45.4 so that the mean is 4.13 (to 2 d.p.)

The deviations of these values from the mean are therefore:

1.17 −0.93 −1.53 −1.23 −0.73 0.57 2.67 −1.23
−0.43 0.97 0.67

The squares of these deviations are:

1.3689 0.8649 2.3409 1.5129 0.5329 0.3249 7.1289
1.5129 0.1849 0.9409 0.4489

The sum of these squared deviations is 17.1619.

Hence the mean of the squared deviations is $^{17.1619}/_{11} = 1.560$ (so that the variance is 1.56 sec^2). The standard deviation must be the square root of 1.560 or 1.249 sec (1.25 seconds to 2 d.p.). Thus, for this distribution, the mean is 4.13 seconds and the standard deviation is 1.25 seconds.

As was indicated earlier, if you have a frequency distribution it is possible to calculate the standard deviation as a by-product of the calculation of the mean; consider the following distribution:

x	4	8	12	16	20	24	28	32	36	40
f	1	7	15	31	22	19	13	6	4	2

The tabulation which appears in Figure 9.4, using an assumed mean x_0 of 16 and a unit size of 4, provides us with the usual form for the calculation of the mean plus one extra column at the end for f'^2 (most easily obtained by multiplying together the two previous columns of d' and fd').

Note that the formula now used for standard deviation is slightly more involved, as might have been expected, but using the values as calculated in the table, it is evaluated straightforwardly. An important advantage gained by this tabulated approach is the use of the assumed mean, which was not used in the previous analysis (of response times); it does make the calculations very much easier. Note also that the data in the example was *not* grouped, thus leading to an *exact* value for the standard deviation; if data *were* to be

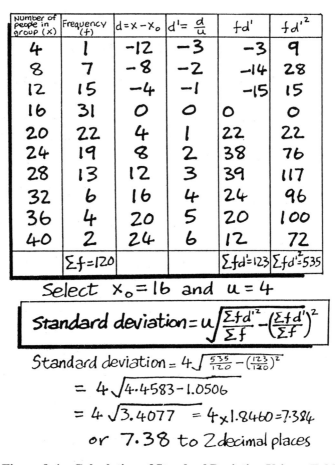

Number of people in group (x)	Frequency (f)	$d = x - x_0$	$d' = \frac{d}{u}$	fd'	fd'^2
4	1	−12	−3	−3	9
8	7	−8	−2	−14	28
12	15	−4	−1	−15	15
16	31	0	0	0	0
20	22	4	1	22	22
24	19	8	2	38	76
28	13	12	3	39	117
32	6	16	4	24	96
36	4	20	5	20	100
40	2	24	6	12	72
	$\Sigma f = 120$			$\Sigma fd' = 123$	$\Sigma fd'^2 = 535$

Select $x_0 = 16$ and $u = 4$

$$\text{Standard deviation} = u\sqrt{\frac{\Sigma fd'^2}{\Sigma f} - \left(\frac{\Sigma fd'}{\Sigma f}\right)^2}$$

$$\text{Standard deviation} = 4\sqrt{\frac{535}{120} - \left(\frac{123}{120}\right)^2}$$

$$= 4\sqrt{4.4583 - 1.0506}$$

$$= 4\sqrt{3.4077} = 4 \times 1.8460 = 7.384$$

$$\text{or } 7.38 \text{ to 2 decimal places}$$

Figure 9.4   Calculation of Standard Deviation Using a Table

grouped then although the mid-interval value for each group serves as the x-value, the calculation can only lead to an approximation for the standard deviation since it fails to take into account how the values are distributed *within* each of the groups.

The formula given for standard deviation involves the factor u; if no division of d-values by u to obtain d'-values is undertaken then it implies that $u = 1$. Furthermore note that in this formula x_0 does not appear at all; thus the assumed mean plays absolutely no part in the calculation of standard deviation even though it is used in calculating the mean. The variance will still be the square of the standard deviation or, in this case, $(7.384)^2 = 54.52$ or 54.5 to 3 sf.

9.6 EXERCISES INVOLVING STANDARD DEVIATION AND VARIANCE

1. The following values are the times, in minutes, taken to run a spreadsheet application on eight different microcomputers:

   6.1   7.3   5.4   6.9   8.7   5.3   6.4   8.3

   Calculate the mean and the standard deviation of these values. State clearly the units involved in each of the results.

   Calculate also the variance of the times, stating clearly the units.

2. Calculate the mean, the standard deviation and the variance of the following set of values of x, the values of orders received (in $) by an independent consultant:

x	50	100	150	200	250	300	350	400	450	500
f	2	7	15	21	16	14	12	8	4	1

3. A survey of the ages of the 200 employees of a company produced the following set of values:

Age (years)	No of employees
16–	12
20–	37
24–	32
28–	24
32–	19
36–	16
40–	18
44–	12
48–	17
52–56	13

Estimate the mean and the standard deviation of the ages.

9.7 SUMMARY

Just as the arithmetic mean is possibly the most useful measure of average so too the standard deviation is widely used to measure dispersion, and both are frequently involved in more sophisticated statistical work.

Although range may not be so widely used as standard deviation it has the great advantage that it can be calculated exceptionally quickly and this has some useful side effects. All too frequently carelessness in calculating standard deviation leads to values being obtained that are wildly inaccurate. One very simple device that works in a large number of cases (but *not* all) is to check that the calculated standard deviation has a value which lies approximately between one eighth and one quarter of the range; if this is true then the standard deviation has at least the correct order of size, whether or not it is arithmetically correct. This test may be used solely for the purpose of establishing that the calculated standard deviation is at least of a size which is appropriate to the data values being analyzed, and so it avoids the putting forward of values which are totally and obviously wrong.

10 Algebraic Terminology and Simple Manipulations

10.1 AN INTRODUCTION TO THE TERMINOLOGY

In programming languages such as Cobol one is familiar with instructions such as:

MULTIPLY HOURS-WORKED BY RATE-OF-PAY GIVING
GROSS-PAY

and experiences very little difficulty in understanding their meaning. However, the instant the same relationship is identified as $G = H * R$ many people express difficulty in understanding it. Both statements mean precisely the same thing, yet in the second we have simplified, not the relationship itself, but the way of writing it, using initial letters to represent the three quantities involved (G for GROSS-PAY, etc) and '*' for 'multiplied by' as well as '=' for 'is equal to' (or 'gives'). This is part of the language of algebra, a neat and convenient way of expressing relationships between quantities; that many people worry about it is due, in part, simply to a lack of familiarity and in part because they do not appreciate that it is no more than a shorthand way of expressing the same relationships that they would be at home with if expressed in flowing text.

Of course, what can be achieved by the use of algebra is a great deal more far-reaching than the foregoing might convey, but the first step has to be a reasonable level of competence with algebraic terminology.

Reverting to the $G = H * R$ statement; this is an equation, as identified by the presence of the 'equals' sign. It tells us that the

153

two quantities are equal to one another, that G, and the product of H with R, both have exactly the same value, or that calculating that product is the way of finding out the value of G. Since H and R will not always have the same values they are called variables and so G must also be a variable. If, on the other hand we looked at the equation $V = 0.15 * T$ where V represents the amount of VAT payable on an amount T, assuming a rate of 15% then whilst V and T are indeed variables the third quantity 0.15 is called a constant since it always has the same value. Had we been asked to provide the same relationship in Cobol it would have been written:

MULTIPLY TOTAL-AMOUNT BY 0.15 GIVING
VAT-PAYABLE

Other symbols are used in algebraic relationships; having already met '*' for 'multiplied by', we can use '+' for 'added to', '−' for 'minus' or 'take away' so that $A = B − C$ can be read as 'A is equal to B minus C' or 'to find A we subtract C from B', etc. Division is signified by '/' which means 'divided by' although 3 divided by 4 might sometimes be denoted as $\frac{3}{4}$, a use of a 'fractional' notation as an alternative to $3/4$ (the 'solidus' notation). In computing the notations invariably used are '*' and '/' for 'multiplied by' and 'divided by' respectively, although this notation is not always used elsewhere. Anyone used to programming in Basic, Fortran or Pascal (or others) will recognise at once the symbols used.

Within the area of terminology will be encountered the use of the word 'power' or 'exponent'. If we write X^4 this may be read as 'X to the power of 4', meaning $X * X * X * X$ so that it is simply a convenient shorthand for expressing the idea of 4 Xs multiplied together and is *not* the same thing as X times 4 (which would be written as $X * 4$ or as $4 * X$). It is rather unfortunate perhaps that the letter X appears so frequently as it does; it is just that, almost by tradition, mathematicians have used it far more than any other letter to represent a variable so that, to many people, 'X' and 'algebra' have become almost synonymous. The use of exponents does also cause us to encounter, on occasions, such things as X^1, which is a less common way of writing just X itself, or X^0 which, whatever the value of the variable X, always has the value 1. Hence:

$$5^3 = 5 * 5 * 5 = 125$$

$$7^4 = 7 * 7 * 7 * 7 = 2401$$

$$3^1 = 3$$

$$6^0 = 1$$

$$(7.9)^1 = 7.9$$

$$(8.326)^0 = 1$$

In some programming languages exponents may not be able to be expressed in the way given here, since all instructions must be 'at line level' so that you may find that X^4 will appear as $X**4$ or as $X\uparrow4$, depending upon the language. Thus treat '**' or '$\uparrow$' as meaning 'to the power of'.

Sometimes there may be a variable such as A which may possibly be used to denote a person's age; provided that we do not use A in the same program to refer to a quite different quantity there is no risk of confusion. However if it was necessary to read in record after record and note the ages of several people, you could scarcely use the same letter A for each one at the same time. In such a case we may use subscripted variables and use A_1 A_2 A_3, etc to denote the various ages (the numbers 1, 2, 3, etc being the subscripts). Again programming languages do not in general permit of 'below the line' values so you may find that $A(1)$ $A(2)$ $A(3)$, etc are used instead of $A_1 A_2 A_3$, etc. This approach is invariably used when we have an array to hold such values; the advantage is that, whilst it separates out the individual values from one another it also retains the idea that each value represents the same quantity, but for each of a number of different people. When programming you will also find that the subscript itself may be variable, eg $A(N)$ to allow for more usable program control structures, thus:

FOR $N = 1$ TO 20

READ $A(N)$

NEXT N

in Basic will read in first the value of $A(1)$ then $A(2)$, etc up to $A(20)$, storing the values into the elements of the array A.

10.2 THE RULES OF ALGEBRA

In algebra which is not to be used in the context of a program, the expression $5f$ is used to denote the product of the constant 5 with the variable f, ignoring the multiplication sign (taking it as being read). Since the computer would attempt to interpret $5f$ as an address instead of as a product the multiplication sign *is* included in expressions used in programming so that it would be referred to instead as $5 * F$ (note that upper case letters are used more often than not). If met with $7g + 5g$, it is easy to appreciate that this is the same thing exactly as $12g$ so that 'seven lots of g' added to 'five lots of g' must simplify to '12 lots of g'. This simplification can be extended so that $15y + 18y - 7y$ reduces to $26y$; in the same way if we had $13xy + 5xy - 7xy$ it simplifies to $11xy$, the fact that one is working with 13 lots of 'the product x times y' does not invalidate the simplification. The important thing to recognise is that you can 'collect up terms' in this way only if dealing with like quantities; if each is an expression in y or if each is an expression in xy then the process is straightforward. However the expression $8x + 7y$ cannot be simplified since one term involves x and the other involves the quite different quantity y; we cannot even simplify $8x^2 - 3x$ since x^2 and x are *different* terms, even if they do have something in common with one another.

Expressions as apparently involved as $8ab^2c$ are only simplified ways of writing '8' times 'a' times 'b squared' times 'c' (note that b squared or b^2 is the same as b to the power of 2, just as b cubed is another way of saying b^3 or b to the power of 3). Hence if $a = 3$, $b = 5$ and $c = 7$ the value of $8ab^2c$ would be 8 times 3 times 25 times 7 or 4200.

Faced with multiplying together expressions such as x^4 and x^3, the result appears as x^7, which is found by adding together the exponents. To appreciate this it is necessary to recognise that $x^4 * x^3$ is really $(x * x * x * x) * (x * x * x)$ or the product of $(4 + 3)$ or $7x$ values together. This cannot be done however with x^5 and y^6 since the exponents must be of like quantities; x^5 multiplied by y^6 can be expressed only as x^5y^6. If the expressions to be multiplied are more involved, such as $3x^2y^3$ times $7x^5y^6$, then this results in $21x^7y^9$ by considering the products of the like terms independently of one another, using the principles described above. If multiplying

$5ab^2c^3$ by $6a^2b^4c$ first remember that a is the same as a^1 and that c is the same as c^1 so that the result is $30a^3b^6c^4$. In the case of $4a^2b$ multiplied by $7b^2c^3$ the result is $28a^2b^3c^3$, since although there is no 'a' term in the second expression there is still effectively a^2 times 1 giving a^2 as the product.

The rules for division are simply the reverse of those for multiplication, so that $a^7/a^2 = a^5$ by subtraction of the exponents.

It can be appreciated that a^7/a^2 is really $\dfrac{a * a * a * a * a * a * a}{a * a}$ and that cancellation leads to the expression $\dfrac{\cancel{a} * \cancel{a} * a * a * a * a * a}{\cancel{a} * \cancel{a}}$ or a^5.

Extending this to more involved cases such as $32a^5b^6/4a^2b^4$ requires the rule to be applied to each part in turn so that the result is $8a^3b^2$ (since $^{32}/_4 = 8$, $a^5/a^2 = a^3$, etc). Again, this can be done only with 'like' terms so that it is not possible to simplify x^6/y^3 in any way. Do not be alarmed if this process leads to negative exponents as in the case of x^3/x^5 which is x^{-2}; the expression x^{-2} is just another way of writing $1/x^2$ or $\dfrac{1}{x^2}$, again another case of algebraic terminology.

Expressions such as $(5x^2y^3)^3$ may be encountered. This is just another way of writing $(5x^2y^3) * (5x^2y^3) * (5x^2y^3)$ so gives the value $125x^6y^9$. Other specific examples like $(x^5)^4 = x^{20}$ and $(y^7)^2 = y^{14}$, and $(a^4)^3 = a^{12}$ illustrate this point further, but may be generalised as $(x^a)^b = x^{ab}$.

Another situation that is likely to be encountered is the use of brackets when multiplying, for example $3a(2a + 5b)$ which is simplified by multiplying first the $2a$ by the $3a$ and then the $5b$ by the $3a$ and adding the results together, thus giving $6a^2 + 15ab$. If wishing to evaluate $5x(4x - 3y)$ the result likewise is $20x^2 - 15xy$. Additional examples are:

$$3p(2p - 6q) = 6p^2 - 18pq$$

$$5x^2(2x - 3y) = 10x^3 - 15x^2y$$

$$7ab(bc + 2a^2 - c) = 7ab^2c + 14a^3b - 7abc$$

You may encounter two (or more) bracketed expressions multiplied together as in the case of $(x + 3)(2x - 5)$; this should be treated as $x(2x - 5) + 3(2x - 5)$; notice that the x and the $+3$ are the two terms from the first bracket. This will now yield $2x^2 - 5x + 6x - 15$ or, since $-5x + 6x$ simplifies to x, the result of the product will now be $2x^2 + x - 15$. It should be remembered at this point that if two numbers are multiplied together the product is positive if *both* numbers are positive *or* if both numbers are negative, but is negative if the two numbers *differ* in sign; thus $-3 * -6 = +18$, $+2 * +5 = +10$, but $-3 * +5 = -15$ and $+2 * -7 = -14$. If the two bracketed expressions being multiplied together are $(3x - 2)$ and $(2x - 7)$ then we get $(3x - 2)(2x - 7) = 3x(2x - 7) - 2(2x - 7)$ which is equal in turn to $6x^2 - 21x - 4x + 14 = 6x^2 - 25x + 14$ as the result.

Further examples are:

$$(x - 6)(2x + 3) = x(2x + 3) - 6(2x + 3)$$
$$= 2x^2 + 3x - 12x - 18 = 2x^2 - 9x - 18.$$

$$(a + 5b)(a - 2b) = a(a - 2b) + 5b(a - 2b)$$
$$= a^2 - 2ab + 5ab - 10b^2 = a^2 + 3ab - 10b^2.$$

$$(x^2 - 3)(2y - 7) = x^2(2y - 7) - 3(2y - 7)$$
$$= 2x^2y - 7x^2 - 6y + 21 \text{ with no further simplification.}$$

A further situation which might be encountered is a bracket raised to a power as in the case of $(2x - 3)^2$, but as this is just another way of writing $(2x - 3)(2x - 3)$ it is calculated to be $4x^2 - 6x - 6x + 9 = 4x^2 - 12x + 9$ straightforwardly.

10.3   EXERCISES

1.   Calculate each of the following:

   (a) 5^4 (e) 49^0

   (b) 3^5 (f) 17^1

   (c) 2^6 (g) 2.3^2

   (d) 8^3 (h) 2.3^0

2.   Simplify as far as possible each of the following:

   (a) $3x + 7x + 15x$

    (b) $6a + 5a - 2a$

    (c) $3x + 8x - 4y$

    (d) $7ab - 2ab + 9ab$

    (e) $8a^2 + 3a - 5a$

    (f) $15a^3 - 7a^3 + 2a^3$

    (g) $16abc - 4abc + 18bc$

3.   Given that $x = 2$, $y = 4$ and $z = 5$, calculate:

    (a) $7xy$

    (b) $2xyz$

    (c) $3x^2z$

    (d) $8xy + 3yz$

    (e) $4x^2yz$

4.   Simplify as far as possible each of the following:

    (a) $3x^2 * 5x^4 * 2x$

    (b) $8a^2b * 3ab^2c$

    (c) $16x^2y^4/4xy^3$

    (d) $81x^2y^6z^4/12x^2y^2z$

    (e) $6a^2b^2c * 5abc * 2b$

    (f) $72a^7b^6c^4/18a^7bc^8$

5.   Given that $x = 3$ and $y = 7$ what are the values of:

    (a) $2xy - x$ (d) $x^2 + y^2$

    (b) $5x$ (e) $(xy)^0$

    (c) $6x^{-1}$ (f) $y^2 + x^3$

6.   Simplify as far as possible each of the following products:

    (a) $4x(2x - 3y)$

    (b) $5a(6a + 3b)$

(c) $6a(a + b) - 3(ab - a^2)$

(d) $(2a + b)(a - 3b)$

(e) $(4x - 1)(2x - 5)$

(f) $(2x + 5)(3x + 1)$

(g) $(3x - 4)^2$

(h) $(4x + y)(4x - y)$

(i) $(3a - 7b)(2a + 3b)$

(j) $(a + 2b)^2 - 5ab$

10.4 THE CONCEPT OF EQUALITY AND EQUATIONS

At the start of this chapter was introduced the idea of an equation as being a way of stating that two expressions have exactly the same value. This is true for $8 * 3 = 24$, for $^{35}/_7 = 5$ or for $18 + 17 - 2 = 33$; it is equally true for $2x + 3 = 19$, even if (at the outset) we do not happen to know what the value of x will be. What is being said is that the value of x has yet to be determined, but that the value of $2x + 3$ will be the same as the value of 19.

Some equations are more involved than this; for example if we say that $4x^2 - 3x = 85$ then, whilst the relationship is not true for most values of x it *is* possible to find two values of x for which it *is* true; the finding of these values or of the one value in the previous equation, is a matter of developing and applying techniques of solution.

Equations range from the simple to the complicated, some have one solution, some have two and some have a very large number of possible solutions. For example, if you consider $x^4 - 2x^3 - x^2 + 2x = 0$, it is true if x is either -1 or 0 or $+1$ or $+2$ (four values in all), but not for any other value of x whatsoever.

It must also be recognised that some equations do not have *any* solutions whatsoever. This may appear to be strange, a heresy almost – surely there *has* to be a solution? – but it is necessary to move away from artificial problems all of which just 'happen' to work out very neatly. Some equations simply do not have any solutions, in precisely the same way that some real-life problems just cannot be solved; what is more, it is even possible to prove that they cannot be solved.

10.5   ELEMENTARY INEQUALITY

For most people their experience of algebra has been of something dealing primarily with the idea of equality, but in fact this presents a very limited aspect of the relationship between two expressions. I might claim that my age was exactly 32 years but there are far more people in the world who are older than 32 and far more who are younger than 32 than ever there are who *are* exactly 32. Inequality is therefore far more common than equality and warrants some investigation.

In dealing with equality the symbol '=' was used to mean 'is equal to'. Now the symbols '>' to mean 'is greater than' and '<' to mean 'is less than' are introduced. Occasionally the symbols '≥' meaning 'is greater than or equal to' and '≤' denoting 'is less than or equal to' will be used.

Statements such as $x > 5$ are called inequations and (like equations) are true for some values of x and not for others. In this very simple case there are an infinity of values of x for which this particular inequation is true, whereas for the 'equivalent' equation $x = 5$ it is true for one value of x only. There may be inequations which are far more involved than this, for example, the statement $8y^2 - 6y < 7$, and in each case, to provide solutions the various techniques which will give rise to these solutions will need to be investigated.

Inequations do occur very frequently, for example in solving management problems, such as those experienced in linear programming or other optimisation techniques. For this reason they often turn up within packages used in the implementation of management information systems.

10.6   RULES OF PRECEDENCE

Consider first the expression $x + y^2$. If aware that x was 3 and that y was 7 we should expect the value of $x + y^2$ to be calculated as 52. However this is because *we* are aware that the 7^2 has to be worked out before the 3 is added; if the calculation had simply been done on a *totally strict* left to right basis the 3 and the 7 would be added, getting 10, before coming to the power 2 which would then yield 10^2 so getting a result of 100. Whilst *we* know this is wrong it is only because we knew to work out the 7^2 before adding the 3. In other words it is accepted that there is a precedence in the order in which the calculations are carried out.

Investigated in this section will be the general rules of precedence that govern the way in which any calculation is carried out in the computer. The stages can be identified as:

(i)   Brackets.

(ii)   Exponents.

(iii)   Multiply and Divide.

(iv)   Add and Subtract.

This provides a priority order. To appreciate just how it works consider the straightforward case of $3x^2 - 7x + 3$ when $x = 1.2$. Note that this would be written as $3*X**2 - 7*X + 3$ for computer use. Hence we are trying to evaluate:

$$3 * 1.2^2 - 7 * 1.2 + 3$$

The first step, since there are no brackets involved, is to work out the exponent, that is to calculate 1.2^2 which is 1.44 so that the current state of the evaluation is:

$$3 * 1.44 - 7 * 1.2 + 3$$

Next is to go to the third level of the priority list which indicates that all multiplications and divisions should be worked out. This is done on a strict left to right basis so that we work out $3 * 1.44$ to give 4.32, hence getting:

$$4.32 - 7 * 1.2 + 3$$

then the second product of $7 * 1.2$ is found to be 8.4, thus giving:

$$4.32 - 8.4 + 3$$

Finally the last level of priority is used to deal with all additions and subtractions again evaluated on the strict left to right basis to give first:

$$-4.08 + 3$$

and finally the result which is:

$$-1.08.$$

The next example is working out $(2x^2 - 7)(x/5 + 6)$ with $x = 4.7$. Note that this time each of the brackets has to be worked out separately before the calculation can proceed. Hence the first bracket is evaluated as:

$$2 * 4.7^2 - 7$$

which, dealing with the exponent first is:

$$2 * 22.09 - 7$$

Then, dealing with the solitary product, produces:

$$44.18 - 7$$

Finally, carrying out the subtraction, the result is:

$$37.18$$

The second bracket is evaluated as:

$$4.7/5 + 6$$

This time, with no exponent, deal first with the division to get:

$$0.94 + 6$$

and then with the addition to obtain the result:

$$6.94$$

Finally, reverting to the original expression and having evaluated both brackets, is found:

$$37.18 * 6.94$$

since the multiplication, even if not stated explicitly, is implied; this gives the result:

$$258.0292.$$

10.7 EXERCISES

1. Use the rules of precedence to evaluate on a step by step basis the expression $3x^3 - 6x^2 - 2x + 12$ when $x = 5$.

2. Evaluate, using the rules of precedence, on a step by step basis, the expression $(4x + 2)(3x - 5)$ with $x = 1.7$.

3. Evaluate, using the rules of precedence, on a step by step basis, the expression $(2x/5 - 6)(x - 7)^2$ with $x = 8.2$.

10.8 MANIPULATION OF EQUATIONS

An equation presents us with two quantities which are known to be equal. Hence, provided that *both sides of the equation are treated in the same way,* anything can be done to them in order that they may be manipulated. Hence equal quantities may be added to both sides or the same quantity subtracted from each side; both may be multiplied by the same thing or each divided by the same (*non-zero*) quantity. Both sides can be squared but this should be done with caution since it introduces extraneous solutions. Thus for example, if one took $x = -7$ and squared both sides the result would be $x^2 = 49$; this is true for $x = -7$ (with which we started) but it is also true for $x = +7$. The second value obtained in this case is the extraneous one and great care needs to be exercised when stating results lest one of them, being extraneous, does not fit the original problem. Here, quite clearly the solution $x = +7$ does not fit in with the original problem which specified $x = -7$.

Manipulation of equations leads to methods of solving equations and will be discussed in greater detail in Chapter 11, but as an example, consider the problem of solving $5x - 12 = 14$.

If you were to add 12 to each side, the equation becomes:

$$5x - 12 + 12 = 14 + 12$$

or $5x = 26$

If you divided both sides by 5 the equation becomes:

$$5x/5 = {}^{26}/_5$$

or $x = 5.2$

which provides the solution, the value of x required.

In order to decide exactly what has to be done to each side, consider the equation $C = \frac{5}{9}(F - 32)$. If we want to 'make F the subject of the equation', that is to get F on its own on one side of the equation so expressing it in terms of C, then the operations that were originally carried out on F to create the right hand side of the equation must be considered. They are:

Subtract 32

Multiply by 5

Divide by 9

Now, reversing each operation (ie multiply instead of divide, etc) and presenting them in the reverse order also gives us:

Multiply by 9

Divide by 5

Add 32

Doing these operations on C, the existing subject of the equation, first of all gives $9C$, then $9C/5$ and finally $9C/5 + 32$ and *this is the formula for F* so that:

$$F = \frac{9C}{5} + 32$$

For a second example of this 'operation' technique, consider $v = u + 32t$ and the desire to 'make t the subject'. The operations performed on t are:

Multiply by 32

Add u

Therefore these must be reversed and carried out in the reverse order on v, the existing subject, hence:

Subtract u

Divide by 32

Thus, starting with v we get firstly $v - u$ then $(v - u)/32$, so that:

$$t = \frac{v - u}{32}$$

One other manipulative technique which can be employed is that of adding two (or more) equations together; this is simply an extension of the idea of adding the same quantity to both sides of an equation. Thus, if $x^2 - 7x = 14$ and $7x + 9 = 15$ then

$$x^2 - 7x + 7x + 9 = 14 + 15$$

or $x^2 + 9 = 29$

This approach can also be used when solving equations and in the next chapter is applied when dealing with simultaneous equations.

10.9   EXERCISES

1. By listing the operations performed on x, make x the subject of the equation

$$y = 7x - 4$$

2. Make u the subject of the equation $s = ut + 16t^2$

3. Add together the equations $5x + 6y = 35$

    and $4x - 6y = 21$

4. Add together the equations $x^2 - 7x + 3 = 0$

    and $7x - 5 = 0$

(xi) $\dfrac{1}{2}(x - 6) + \dfrac{1}{3}(2x + 4) = \dfrac{1}{4}(x - 3)$

(xii) $\dfrac{1}{5}(2x - 6) - \dfrac{1}{2}(x + 3) = 7$

11.3 AN INTRODUCTION TO SIMULTANEOUS EQUATIONS

The equation $3x = 12$ has the unique solution given by $x = 4$. However, if given the equation $3x + 4y = 12$, involving the two variables x and y, then any solution must state both a value for x as well as a value for y. In this case $x = 4$, $y = 0$ provides an adequate solution, but so too does $x = 2$, $y = 1.5$ or $x = 0$, $y = 3$, or $x = 12$, $y = -6$, etc. In other words one linear equation in the two variables x and y possesses an infinite number of equally valid solutions. A second linear equation in x and y such as $5x + 12y = 4$ also has an *infinity* of solutions such as $x = 11$, $y = -4.25$ or $x = 14$, $y = -5.5$, etc. However, even if each equation on its own has an infinity of solutions it is possible (in the majority of cases) to find one solution which is *common to both*, in this case $x = 8$, $y = -3$. This solution, abbreviated to $(8, -3)$ is described as being *the solution* to the pair of linear simultaneous equations:

$$3x + 4y = 12$$
$$5x + 12y = 4$$

Hence there is a need to explore techniques which can be used to obtain this unique solution of such a pair of equations. Two techniques will be investigated; the first using a totally algebraic approach and the second using graphical methods.

The examples will be restricted to equations in two variables (such as x and y) which are themselves linear and do not involve such terms as x^2, xy and y^2, etc.

11.4 SOLUTION OF LINEAR SIMULTANEOUS EQUATIONS BY ALGEBRAIC MEANS

Consider the pair of equations:

$$6x + 7y = 44$$
$$4x - 7y = 6$$

which shall be called equations (i) and (ii) respectively.

Note at once that the term in y is the same in each case, though with different signs. This is a particularly convenient situation since adding equations (i) and (ii) together will produce $10x = 50$ with the y terms vanishing totally. With no y term now present the x value is immediately found to be $^{50}/_{10}$ or $x = 5$. Returning with this value to equation (i) and substituting it for x produces $30 + 7y = 44$, from which we derive $7y = 44 - 30 = 14$ whence we get $y = 2$. It is necessary to check that the solution is correct so substitute $x = 5$, $y = 2$ into equation (ii) to find the left-hand side is $20 - 14$ or 6 which agrees with the right-hand side so confirming the solution. It is essential not to check the solution with the same equation as was used to find the value of y. Equation (ii) could have been used in which to substitute $x = 5$ and to find the value of y whilst reversing the checking procedure by using equation (i) would have been just as effective.

In the previous example $7y$ was in both equations (i) and (ii) but with different signs; where there are two equal terms having the same sign these can be dealt with by subtracting rather than by adding the equations. Suppose we had:

$$4x + 9y = 3 \qquad \text{(i)}$$
$$4x + 2y = 10 \qquad \text{(ii)}$$

Both x terms are equal and have the same sign. Subtraction produces:

$$7y = -7$$

whence $y = -1$. Substitution of $y = -1$ into equation (ii) produces $4x - 2 = 10$ or $4x = 12$ so that $x = 3$. Checking the $x = 3$, $y = -1$ value into equation (i) produces a left-hand side of $12 - 9$ or 3 which matches the right-hand side, so confirming the solution of the pair of simultaneous equations to be $(3, -1)$. Note it is important to check the result since there are several stages in the process of obtaining the solution where errors could quite easily have occurred. It is equally true that using the same equation for checking purposes as was used for the substitution gives a fallacious confirmation of correctness.

Consider next the case in which two equal terms are *not* present.

By manipulation of the two equations it can be ensured that two equal terms can be made to appear allowing the pair of equations to be dealt with in precisely the manner used for the two previous examples. Consider the case of:

$$3x - 2y = 10 \qquad \text{(i)}$$

$$4x + 5y = 21 \qquad \text{(ii)}$$

Either the x terms or the y terms can be made equal; if the latter option is chosen this can be achieved by multiplying equation (i) by 5 and equation (ii) by 2, thus producing:

$$15x - 10y = 50$$

$$8x + 10y = 42$$

Addition yields $23x = 92$ so that $x = {}^{92}/_{23} = 4$. Substitution of $x = 4$ into the original equation (i) gives $12 - 2y = 10$ so that $-2y = 10 - 12 = -2$ or $y = 1$. Check by substituting $x = 4$, $y = 1$ into the original equation (ii), in which case the left-hand side reduces to $16 + 5$ which is 21, the same as on the right-hand side. Hence the solution is (4,1).

The decision to multiply equations (i) and (ii) by 5 and 2 respectively was made by considering the LCM of the initial coefficients of y, namely 2 and 5, which is 10, and then considering how each should be multiplied so as to produce 10. Just as easily, the x terms could have been made equal. Since the coefficients were 3 and 4, their LCM is 12. Hence equation (i) would have been multiplied by 4 and equation (ii) by 3 to give:

$$12x - 8y = 40$$

$$12x + 15y = 63$$

This time, subtraction is needed to give $-23y = -23$ from which we get $y = 1$. The x value can be obtained and the solutions checked as before.

For one final example consider the equations:

$$3x + 10y = 44 \qquad \text{(i)}$$

$$7x - 15y = -89 \qquad \text{(ii)}$$

Since the LCM of 10 and 15 is 30, multiply equation (i) by 3 and equation (ii) by 2 to produce:

$$9x + 30y = 132$$

$$14x - 30y = -178$$

Addition yields $23x = -46$ so that $x = -2$. Substitution of $x = -2$ into the equation (i) gives $-6 + 10y = 44$ or $10y = 50$, so that $y = 5$. Checking the result by substituting $x = -2$, $y = 5$ into equation (ii) gives a left-hand side of $-14 - 75$ or -89, confirmed by the right-hand side. The solution is therefore $(-2, 5)$.

11.5 SOLUTION OF LINEAR SIMULTANEOUS EQUATIONS BY GRAPHICAL METHODS

For the sake of an example, consider the pair of equations:

$$2x + 5y = 25 \qquad \text{(i)}$$

$$5x - 3y = 16 \qquad \text{(ii)}$$

Each of these equations is, independently, that of a straight line drawn on standard graph paper. In order to appreciate this consider equation (i). The values $(0, 5)$, $(5, 3)$, $(10, 1)$, $(15, -1)$, etc all satisfy this equation and the points with these as their coordinates can *all* be plotted; in doing so they will be seen to lie on a straight line. Whilst it is sufficient to plot three points only to draw a straight line, if any value of x whatsoever is taken, the corresponding value of y can be calculated and the pair plotted to confirm that the point does indeed lie upon the line as stated.

Considering equation (ii) it can be satisfied by a different set of points such as $(2, -2)$, $(5, 3)$, $(8, 8)$, $(11, 13)$, $(14, 18)$, etc. Both lines are plotted on the graph in Figure 11.1. The point where the lines cross must lie on both lines, indeed will be the only point to do so; as such its coordinates must be the (x, y) values that satisfy both equations and so must define the solution of the given pair of equations.

By reference to the graph it follows that $x = 5$, $y = 3$ must be the solution of the original pair of equations.

Although this approach is far more laborious than the algebraic

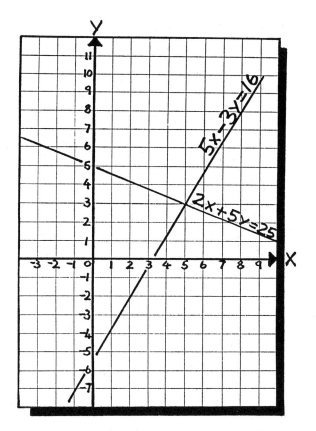

**Figure 11.1   Solving Linear Simultaneous Equations by Use of
Graphs**

method when dealing with linear equations, its virtue is that it can
be used no matter whether the equations are linear or not. If, for
example, the equations had been the non-linear:

$$x^3 - 6y^2 = 5xy$$

$$x^2 + 7y = xy^2$$

then an algebraic solution would be excessively clumsy, and
probably impossible, yet the graphical approach allows for a
reasonably quick method of solution, provided that it is realised

that in such cases there will almost certainly be several points of intersection since such equations rarely have a unique solution. However, there may be some loss of accuracy with the graphical method since it depends very heavily upon the drawing accuracy of the user.

11.6   EXERCISES INVOLVING SIMULTANEOUS EQUATIONS

1.   Use algebraic methods to solve:

$$3x - 5y = 22$$
$$7x + 5y = 18$$

2.   Solve algebraically:

$$8x + 3y = 46$$
$$5x + 7y = 39$$

3.   Solve by algebraic methods:

$$3x - 5y = 22$$
$$7x + 5y = 18$$

4.   Draw the graphs of $2x + 3y = 8$ and of $4x - y = 9$ on graph paper. Hence find the solution of:

$$2x + 3y = 8$$
$$4x - y = 9$$

as accurately as possible.

11.7   INTRODUCTION TO QUADRATIC EQUATIONS

A quadratic equation in x is one which contains terms in x^2, x and a constant. As such the following are examples of quadratic equations in x:

$$5x^2 - 3x + 7 = 0$$
$$x^2 + 12x - 16 = 0$$

However it is possible for the x term to be absent *or* for the constant term to be missing provided that the x^2 term is still present, so that the following are also classified as quadratic equations:

$$5x^2 + 4x = 0$$

$$6x^2 - 13 = 0$$

To solve a quadratic equation is a matter of finding which *two x* values satisfy the equation. In a linear equation there will always be *one x* value but in a quadratic equation there are always two. For example, consider the equation:

$$x^2 - 5x + 6 = 0$$

this is satisfied *either* by $x = 2$ *or* by $x = 3$ but by no other value of x. It must also be noted that the earlier examples of quadratic equations were after simplification so $(x + 3)(x - 7) = 5x(2x - 3)$ is also a quadratic equation but only by multiplying out the brackets to get

$$x^2 + 3x - 7x - 21 = 10x^2 - 15x$$

and collecting up terms to get

$$x^2 - 10x^2 - 4x + 15x - 21 = 0$$

or $$-9x^2 + 11x - 21 = 0$$

will this become evident. Never therefore attempt to classify an equation as linear or as quadratic (or as anything else) until it has been simplified as far as it will go.

There are three prime methods for solving quadratic equations, by factorisation, by use of 'the formula' and by graph. Of these the graphical approach is rarely used since it tends to lack accuracy and is time-consuming and slower than the other methods. The graphical approach is dealt with more fully in Chapters 12 and 13.

11.8 SOLUTION OF QUADRATIC EQUATIONS BY FACTORISATION

If asked to multiply together $(3x - 2)$ and $(x + 5)$ we would get $3x^2 + 13x - 10$, and if asked to solve the equation $3x^2 + 13x - 10 = 0$ (which is quite clearly a quadratic equation in x), this could be rewritten as $(3x - 2)(x + 5) = 0$ since we *already* know that the quadratic expression $3x^2 + 13x - 10$ can be expressed as the product of the two linear expressions $3x - 2$ and $x + 5$.

Now, if two quantities are multiplied together and produce 0 then one or other (or both) of the quantities must itself be 0. Thus either $3x - 2 = 0$ or $x + 5 = 0$; in the former case $x = 2/3$ and in the latter case $x = -5$. Hence $x = -5$ or $x = 2/3$ must be the two values of x which satisfy the original quadratic equation.

To check these results try substituting the two values into the left-hand side of the original equation:

$$3(-5)^2 + 13(-5) - 10 = 3(25) - 65 - 10$$

$$= 75 - 65 - 10 = 0 \text{ so that } x = -5 \text{ satisfies the equation.}$$

Equally　$3(2/3)^2 + 13(2/3) - 10 = 3(4/9) + 26/3 - 10$

$$= 4/3 + 26/3 - 10 = 10 - 10 = 0$$

so that $x = 2/3$ also satisfies the equation.

It follows therefore that if the linear factors of the quadratic expression can be obtained (in the example, $3x^2 + 13x - 10$), then we can solve the equation quite straightforwardly (which, in this case is $3x^2 + 13x - 10 = 0$) by setting each linear expression equal to 0 and solving the two linear equations. Unfortunately, not all quadratic equations do factorise but for those that do, the technique outlined below will always work.

If considering the general quadratic equation as $ax^2 + bx + c = 0$ in which a, b and c have constant values (so in the previous case $a = 3$, $b = 13$ and $c = -10$) then first the product ac has to be found. The next step is to list all the possible pairs of factors which give ac, so that if $ac = 12$ we would list $(1, 12)$ $(2, 6)$ $(3, 4)$ $(-1, -12)$ $(-2, -6)$ $(-3, -4)$ which gives us all such pairs (integer values only). Then, *from this list* select the pair which add together to give b. Using the $ac = 12$ example and given that b is 7, the pair would have to be $(3, 4)$ since not one of the other pairs add up to give 7. At this point the equation is rewritten but replacing the bx term, so if we did indeed have $7x$, it would be rewritten as $3x + 43x$, using the factor pair selected. The process is shown here for the equation $6x^2 + 7x + 2 = 0$ (in which $a = 6$, $b = 7$ and $c = 2$), so that ac is indeed 12 and b is indeed 7:

$$6x^2 + 3x + 4x + 2 = 0$$

Now factorise the left-hand side taking the first two terms together and then the second two terms together to produce:

$$3x(2x + 1) + 2(2x + 1) = 0$$

A common factor, $2x + 1$, has emerged so the equation can now be written as:

$$(2x + 1)(3x + 2) = 0$$

from which either $2x + 1 = 0$ or $3x + 2 = 0$, hence we shall obtain the two solutions as $x = -^1/_2$ or $x = {}^{-2}/_3$.

The example given in Figure 11.2 shows a simpler equation but uses exactly the same method.

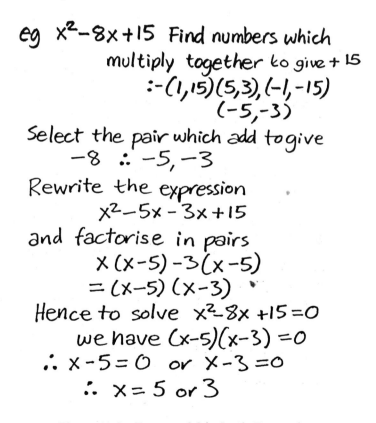

eg $x^2 - 8x + 15$ Find numbers which
multiply together to give + 15
:- (1, 15) (5, 3), (-1, -15)
(-5, -3)
Select the pair which add to give
-8 ∴ -5, -3
Rewrite the expression
$x^2 - 5x - 3x + 15$
and factorise in pairs
$x(x-5) - 3(x-5)$
$= (x-5)(x-3)$
Hence to solve $x^2 - 8x + 15 = 0$
we have $(x-5)(x-3) = 0$
∴ $x - 5 = 0$ or $x - 3 = 0$
∴ $x = 5$ or 3

Figure 11.2   Factors of Quadratic Expressions

In this example we have $a = 1$, $b = -8$ and $c = 15$ so that the product ac is 15; the rest follows from the earlier example directly.

The next example given in Figure 11.3 is slightly longer because it uses $a = 3$, $b = -19$ and $c = -14$ so the factors of ac (ie of -42) take rather longer to list and it also takes longer to find which pair add up to the value of b, but the process is identical to that used on the first two examples.

$$3x^2 - 19x - 14 \text{ Factors of 3 times}$$
$$-14 \text{ ie } -42$$
$$\text{Are:- } (1,-42)(-1,42)(2,-21)$$
$$(-2,21)\ (3,-14)\ (-3,14),(6,-7)$$
$$(-6,7)$$
$$\text{Those which add to give}$$
$$-19 \text{ are } 2,-21$$
$$\therefore 3x^2 + 2x - 21x - 14$$
$$= x(3x+2) - 7(3x+2)$$
$$= (3x+2)(x-7)$$
$$\therefore \text{To solve } 3x^2 - 19x - 14 = 0$$
$$\text{We have } (3x+2)(x-7) = 0$$
$$\therefore x = -\tfrac{2}{3} \text{ or } 7$$

Figure 11.3   Example of Solving a Quadratic Equation

Note that a quadratic equation in a single variable will always have either:

2 different solutions (as in each example so far seen); or

2 equal solutions; or

no real solutions at all.

The '2 equal solutions' case is exemplified by $x^2 - 6x + 9 = 0$ which can be factorised to give $(x - 3)(x - 3) = 0$ so that $x = 3$ or $x = 3$, ie the same value twice over. Occasionally the case of 'no real solutions' may be encountered. This is dealt with more fully in section 11.9.

It should also be recognised that if factors of ac that add together to yield b cannot be obtained then that particular equation is not capable of being factorised in which case an alternative method of solution is used, usually being that by formula.

11.9   SOLUTION OF QUADRATIC EQUATIONS BY FORMULA

Although not all quadratic equations can be factorised one can always revert to the use of the formula which is quoted in Figure 11.4 below.

$$ax^2 + bx + c = 0$$

Solutions are $x = \dfrac{-b \pm \sqrt{b^2 - 4ac}}{2a}$

eg. $5x^2 - 8x - 3 = 0$

$\therefore a = 5 \quad b = -8 \quad c = -3$

Solutions are $x = \dfrac{-(-8) \pm \sqrt{(-8)^2 - 4(5)(-3)}}{2(5)}$

$= \dfrac{8 \pm \sqrt{64 + 60}}{10}$

$= \dfrac{8 \pm \sqrt{124}}{10}$

$= \dfrac{8 + 11 \cdot 14}{10} \quad \text{or} \quad \dfrac{8 - 11 \cdot 14}{10}$

$= \dfrac{19 \cdot 14}{10} \quad \text{or} \quad \dfrac{-3 \cdot 14}{10}$

$= 1 \cdot 914 \quad \text{or} \quad -0 \cdot 314 \ (\text{To 3d.p})$

Figure 11.4   Formula for Solution of Quadratic Equations

The process is, of course, quite automatic. However, considerable care must be taken because, especially in respect of the signs, there are plenty of areas where errors can be created by carelessness. For the sake of an example, consider the solution of $x^2 + 3x - 11 = 0$. In this case we have $a = 1$, $b = 3$ and $c = -11$, so that the solutions are:

$$x = \frac{-(3) \pm \sqrt{(3)^2 - 4(1)(-11)}}{2(1)}$$

$$= \frac{-3 \pm \sqrt{9 + 44}}{2}$$

$$= \frac{-3 \pm \sqrt{53}}{2}$$

$$= \frac{-3 + 7.280}{2} \quad \text{or} \quad \frac{-3 - 7.280}{2}$$

$$= \frac{4.280}{2} \quad \text{or} \quad \frac{-10.280}{2}$$

$$= 2.14 \text{ or } -5.14 \text{ to 3 sf in each case.}$$

The section on solution by factorisation indicated that there were three different possible outcomes to solving a quadratic equation. The 'formula' method can indicate for us immediately which of the three categories will be appropriate for a given quadratic equation:

If $b^2 > 4ac$ we get two different solutions.

If $b^2 = 4ac$ we get two equal solutions.

If $b^2 < 4ac$ we get no real solutions at all.

Hence, merely be examining the original equation and extracting the values of a, b and c and by testing the relationship between b^2 and $4ac$ the category to which the equation belongs can be found.

Thus, for $x^2 - 8x + 16 = 0$ we have $a = 1$, $b = -8$ and $c = 16$ so that b^2 is 64 and the value of $4ac$ is $4(1)(16) = 64$. Since $b^2 = 4ac$ this equation will have two equal solutions.

If we have the equation $3x^2 + 2x + 15 = 0$ then $a = 3$, $b = 2$ and $c = 15$ so that b^2 is 4 and $4ac$ is $4(3)(15) = 180$; in this case $b^2 < 4ac$ which means that this equation *cannot be solved* (in the terms of x values that lie currently within our experience). Do note that when applying the formula to the solution of a quadratic equation you may come across a negative value inside the square root *once the value of $b^2 - 4ac$ has been worked out;* this means that no solutions exist to the equation so it is not possible to proceed any further.

The accuracy of this method is dependent solely upon the accuracy with which calculations such as the calculation of the square root and divisions, etc can be performed, so it depends entirely upon the capacity of the electronic calculator used, or upon the slide rules, the tables, or the computer in calculator mode.

11.10 EXERCISES IN SOLVING QUADRATIC EQUATIONS

1. Use the factorisation method to solve each of the following:

   (a) $x^2 + 6x + 5 = 0$

   (b) $x^2 + 2x - 24 = 0$

   (c) $2x^2 + 7x + 3 = 0$

   (d) $3x^2 - 8x + 5 = 0$

   (e) $6x^2 + 7x + 2 = 0$

   (f) $12x^2 + 17x - 5 = 0$

   (g) $15x^2 + x - 2 = 0$

   (h) $30x^2 + x - 20 = 0$

2. Use the formula to solve, in each case to 3 sf:

   (a) $x^2 - 2x - 3 = 0$

   (b) $x^2 + 7x - 2 = 0$

   (c) $2x^2 + 9x - 7 = 0$

    (d) $3x^2 - 6x + 2 = 0$

    (e) $5x^2 - 8x - 1 = 0$

3.   Determine, for each of the following quadratic equations, whether they have (i) 2 different solutions;

                                   (ii) 2 equal solutions;

                                   (iii) no real solutions.

    (a) $3x^2 - 4x + 2 = 0$

    (b) $x^2 - 10x + 25 = 0$

    (c) $2x^2 - 7x + 1 = 0$

    (d) $4x^2 + 2x - 3 = 0$

    (e) $4x^2 - 20x + 25 = 0$

11.11   SOLUTION OF SIMPLE INEQUATIONS

The concept of the inequation was raised in Chapter 10 and, as they often appear in the context of linear programming (a very commonly-used management technique), so some simple treatment of their manner of solution is appropriate.

   The rules governing the treatment of inequations are very similar indeed to the set of rules used for dealing with equations but with some modifications. Equal quantities can be added or subtracted to or from both sides of an inequation and both sides can also be multiplied by the same *positive* quantity. Hence, if we have $x + 4 > 2$ we can also have $x > 2 - 4$ or $x > -2$; equally well if we have $3x > 6$ then $x > 2$. However both sides of an inequation may be multiplied or divided by a *negative* quantity only if the sign of the inequation is changed in the process. Hence whilst $-2x > 7$, dividing both sides by -2 gives $x < -3.5$; furthermore if we had the inequation $\dfrac{x}{3} < -5$ then if it was multiplied by -3 we get $-x > 15$.

   Unequal quantities may be added to the two sides of an inequation provided that the larger quantity is added to the side which is already the larger. Hence if $x > y$ we can also have $x + 4 > y + 3$.

   The examples in Figure 11.5 illustrate both the solution of very simple inequations and also the way that the solutions can be

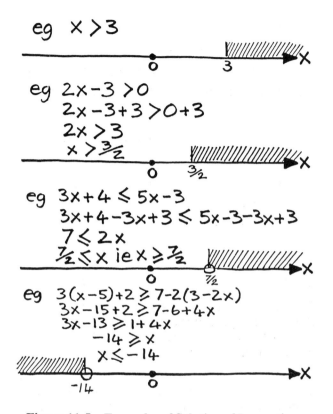

eg $x > 3$

eg $2x - 3 > 0$
$2x - 3 + 3 > 0 + 3$
$2x > 3$
$x > \frac{3}{2}$

eg $3x + 4 \leqslant 5x - 3$
$3x + 4 - 3x + 3 \leqslant 5x - 3 - 3x + 3$
$7 \leqslant 2x$
$\frac{7}{2} \leqslant x \ ie \ x \geqslant \frac{7}{2}$

eg $3(x-5) + 2 \geqslant 7 - 2(3 - 2x)$
$3x - 15 + 2 \geqslant 7 - 6 + 4x$
$3x - 13 \geqslant 1 + 4x$
$-14 \geqslant x$
$x \leqslant -14$

Figure 11.5   Examples of Solution of Inequations

represented on a number line; note that in the first case, $x > 3$, the
shaded area indicating the solution set, does *not* include $x = 3$.
However in the third case $x \geqslant 3.5$ the value $x = 3.5$ is circled to
indicate that it is included in the solution set.

12 Plotting and Sketching Graphs

12.1 FUNCTION NOTATION AND ITS USE

Having already become used to writing down statements such as $y = 2x^2 - 7x + 3$ and classifying them as equations, it is now worth considering an alternative way of treating expressions such as those on the right-hand side of such an equation. Any expression, be it $2x^2 - 7x + 3$ or $5x + 14$ or $\dfrac{2}{x + 2}$, etc is a function of x, in that its value depends solely upon the value assigned to x. Therefore, rather than use the '$y = $ ' approach we can instead simply state that the expression is defined to be a particular function of x, denoted by $f(x)$, hence $f(x) = 2x^2 - 7x + 3$.

This has an immediate advantage in that it enables us to use $f(2)$ to mean the value of $f(x)$, when x is replaced by 2. Hence $f(2) = 2(2)^2 - 7(2) + 3$ or $f(2) = 8 - 14 + 3 = -3$. Likewise $f(1) = 2(1)^2 - 7(1) + 3 = 2 - 7 + 3 = -2$ and also that $f(4) = 2(4)^2 - 7(4) + 3 = 32 - 28 + 3 = 7$. The letter f stands for 'function' and once a particular function has been specified by defining an algebraic relationship in which $f(x) = $ 'something', then the function f itself can be discussed by referring to the values it takes for different valus of x and so on; this leads us to discuss the 'behaviour' of f as x is allowed to vary.

There may be a need to refer to a number of different functions in which case it is normal to use a number of different letters other than f. Hence $f(x)$ may well represent one function, whilst $g(x)$ and $h(x)$ represent two other functions. Thus we could have $f(x) = 2x + 7$, $g(x) = 3x^2 - 2x + 6$ and new functions can be created by combining the ones so far defined; hence it is in order to define $f(x) + g(x)$ as $2x + 7 + 3x^2 - 2x + 6 = 3x^2 + 13$ which produces a new function in terms of others. We may define $h(x) = 3x^2 + 13$ or say that $h(x) = f(x) + g(x)$.

12.2   EXERCISES

1.   Given that $f(x) = 3x^2 - 2x + 5$ what is:

   (i) $f(0)$

   (ii) $f(2)$

   (iii) $f(-1)$?

2.   Given that $g(x) = \dfrac{x + 5}{2x - 1}$ what is:

   (i) $g(2)$

   (ii) $g(-4)$?

3.   If $f(x) = 6x^2 - 3x + 2$ and $g(x) = 7x + 5$, what is:

   (i) $f(x) + g(x)$

   (ii) $f(x) - g(x)$

   (iii) $2g(x)$?

12.3   WHAT IS A GRAPH?

Having introduced the idea of a function a graph may now be described as being a visual method of illustrating the behaviour of a particular function. The graph is able to show easily how, as values of the variable x change, the value of $f(x)$ is changing; for example, does $f(x)$ change slowly or rapidly? Is it increasing or decreasing as x increases and decreases? The graph, by providing visual impact, is far more easily 'digested' for many people than the production of a table of values of $f(x)$ for various values of x, or indeed more readily understood than the statement that '$f(x) = \ldots$' when we define the function by means of an algebraic statement. The latter approach is really only suitable for those mathematicians who can easily appreciate the nature of the function by recognising the form of its $f(x)$ definition.

The graph is able to indicate for what values of x the function reaches its highest or its lowest values, if indeed it has any; it shows where specific values of $f(x)$ are obtained and it shows when the function is increasing or decreasing as x increases. Likewise it can show when two functions have the same value, an important characteristic exploited in Chapter 11 in the graphical solution of simultaneous equations.

On occasion, the graph may prove to be the only way to describe some function at all. For example, if a monitoring device is used to record a patient's heartbeat we will obtain a trace on a screen, which *is* a graph, but there is unlikely to be any way in which an equation could be produced which would determine the heartbeat at any particular instant. In such cases the graph is not just an alternative way of representing a function; it may indeed be the *only* way.

12.4   RECTANGULAR CO-ORDINATES

Having established *why* a graph needs to be used, the next stage is to find out *how*. To do this we need to be able to record both x and $f(x)$ for a very wide number of values of x in the same diagram; these two values will be referred to as the co-ordinates of a point and in the diagram the function will be represented by a series of points, linking up the points on a flat surface (since we will confine ourselves to work in two dimensions) to form a smooth curve, the graph of the function.

The most common method uses what are technically called *rectangular Cartesian co-ordinate axes;* which will simply be referred to as the axes. The system was first developed for use in the seventeenth century by René Descartes, the French philosopher-mathematician, hence the 'Cartesian' in the title. The diagram consists of two straight line axes which are at right-angles to one another (hence 'rectangular') and which cross at a point called the origin. These two lines are usually referred to as the x-axis and the y-axis respectively, as illustrated in the diagram given in Figure 12.1.

Note first that points are marked along the (horizontal) x-axis, in equal steps, with 0 at the origin and positive values to the right,

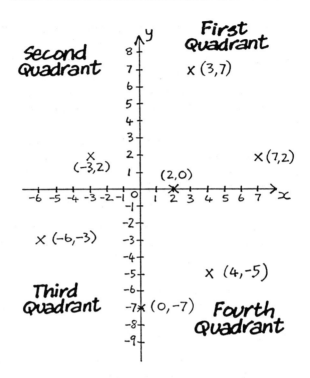

Figure 12.1   Rectangular Cartesian Co-ordinate Axes

negative values going to the left. It is essential that successive points do represent equal numeric intervals, so 0, 1, 2, 3, 4, etc is the most common system; it follows that intermediate values such as 1.5, although not actually marked, will be half-way between the 1 and the 2 values. In a similar manner the (vertical) y-axis has positive values going upwards from 0 at the origin and negative values downwards; the other principles apply equally well to the y-values.

The area of the diagram is subdivided into four *quadrants* by the two axes; within the first quadrant all x-values and all y-values are positive, which can be checked by looking at the two parts of the axes which provide the boundaries of the quadrant. In the second quadrant, all x-values are now negative and all y-values are positive and so on in each quadrant.

It is now possible to position a point so as to reflect its x and $f(x)$ values; it is referred to as having co-ordinates $(x, f(x))$ and to find its position go to the x-value on the x-axis and then move vertically up or down to the value of $f(x)$ on the y-axis. Thus, in the diagram, seven points are plotted to illustrate the technique; for example the point $(3,7)$ was plotted by going to 3 on the x-axis and then going vertically upwards until level with 7 on the y-axis and marking the point with a cross. Likewise $(-3, 2)$ was found by going to -3 on the x-axis and vertically upwards until level with 2 on the y-axis; $(-6, -3)$ required us to go vertically downwards from -6 on the x-axis so as to get level with the negative y-value of -3. As a result each point is uniquely defined in the system.

Other co-ordinate systems do exist, each with its own particular uses, such as one which uses oblique axes, which do not cross at right-angles. Another system uses polar co-ordinates, with a combination of distances and angles to define each point uniquely. This book however, does not cover them in more detail but is restricted to rectangular axes as introduced in this section.

12.5   PRODUCING TABLES OF VALUES

To be able to draw the graph of a function a set of points needs to be plotted and this in turn means that for a range of values of x the values of $f(x)$ must also be calculated. This can be done manually or by a very simple program. The manual method is illustrated in the diagram shown in Figure 12.2 in which the function $f(x) = 3x^2 - 8x + 4$ is investigated.

x	-3	-2	-1	0	1	2	3	4	5
x^2	9	4	1	0	1	4	9	16	25
$3x^2$	27	12	3	0	3	12	27	48	75
$-8x$	24	16	8	0	-8	-16	-24	-32	-40
4	4	4	4	4	4	4	4	4	4
$f(x)$	55	32	15	4	-1	0	7	20	39

Figure 12.2   Manual Method for Plotting a Range of Values

Note that in this case the x values -3 to 5 are listed along the top line since it is the intention to produce the graph of $f(x)$ only for values of x from -3 to 5 inclusive. Since $f(x)$ will include a term in x^2 the corresponding x^2 values are listed on the second line. The 'double-bar' separates lines 1 and 2 from lines 3, 4 and 5 and it is in these next three lines that the various parts of the expression that go to make up $f(x)$ are listed and calculated. The first line gives the values of $3x^2$ for each value of x, calculated all the more easily since the x^2 values were worked out earlier. The second line gives the equivalent values of $-8x$ and the third repeats the constant value $+4$. Since $f(x)$ is defined to be the sum of these three terms they are then added up on a column by column basis and the $f(x)$ values listed on the bottom line; note that the x and x^2 lines are *not* incorporated into this addition. Thus, for each value of x in the range -3 to 5 inclusive the $f(x)$ value has been calculated. By use of computer the same values could easily be calculated using a simple program such as the three lines of Basic code which appear in Figure 12.3.

```
7Ø FOR X = -3 TO 5
8Ø PRINT X, 3*X*X-8*X+4
9Ø NEXT X
```

Figure 12.3   Code in Basic for Calculation of Values

A second example follows in which the graph of $f(x) = \dfrac{2x + 3}{x + 6}$ has to be drawn for values of x from -4 to 6 inclusive. In Table 12.1 the values of $2x + 3$ and $x + 6$ are calculated and then divided to produce the $f(x)$ line.

x	-4	-3	-2	-1	0	1	2	3	4	5	6
$2x+3$	-5	-3	-1	1	3	5	7	9	11	13	15
$x+6$	2	3	4	5	6	7	8	9	10	11	12
$f(x)$	-2.5	-1	-0.25	0.2	0.5	0.71	0.88	1	1.1	1.18	1.25

Table 12.1 Tabulation to Produce a Graph of $f(x) = \dfrac{2x+3}{x+6}$

Equally well this could have been provided by the Basic code:

```
$
110 FOR X = −4 TO 6
120 PRINT X, (2*X + 3)/(X + 6)
130 NEXT X
$
```

Finally, if one had to produce a graph of $f(x) = 2x^3 - 7x^2 + 6x + 5$ for x values from -2 to 7 inclusive the tabulation could be done like that shown in Table 12.2.

x	-2	-1	0	1	2	3	4	5	6	7
x^2	4	1	0	1	4	9	16	25	36	49
x^3	-8	-1	0	1	8	27	64	125	216	343
$2x^3$	-16	-2	0	2	16	54	128	250	432	686
$-7x^2$	-28	-7	0	-7	-28	-63	-112	-175	-252	-343
$+6x$	-12	-6	0	6	12	18	24	30	36	42
$+5$	5	5	5	5	5	5	5	5	5	5
$f(x)$	-51	-10	5	6	5	14	45	110	221	390

Table 12.2 Tabulation to Produce the Graph of
$$f(x) = 2x^3 - 7x^2 + 6x + 5$$

12.6   PLOTTING THE GRAPH

Once the table of values is available, the points with co-ordinates $(x, f(x))$ have to be plotted on the rectangular axes. The choice of scale on both axes will be influenced by the range of x values and by the range of $f(x)$ values. If one considers the $f(x) = 3x^2 - 8x + 4$ function introduced in section 12.5, the x values go from -3 to 5 whilst the $f(x)$ values, which have to be plotted on the y-axis, range from -1 to 55. Hence, whilst a scale of 1 cm to 1 unit may be suitable for the x-axis as used below, the y-axis will require a scale of, for example 1 cm to 10 units to enable the graph to be drawn within the physical limitations of the page (Figure 12.4).

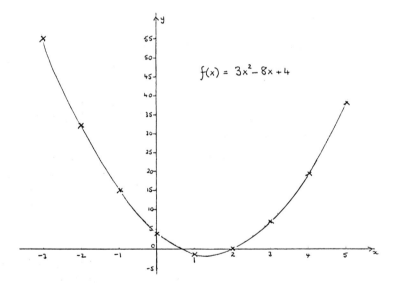

Figure 12.4   Plotting a Graph from Tabulated Data

With the points plotted, a smooth curve has to be drawn to go through all the points; this is not always easy to do, requiring a steady hand or sometimes the use of a flexible plastic curve to ensure that the line so drawn really is smooth.

Having got the curve it could now be used for one of a variety of tasks such as:

(i) By going vertically upwards from the point 3.6 on the x-axis until the graph is cut we can find the value of $f(3.6)$ by reading horizontally onto the y-axis.

(ii) By going horizontally from the point 25 on the y-axis to meet the graph (in two points, one to the left of the y-axis and the other to its right) we could read vertically down onto the x-axis to read off the x-values which make $f(x) = 25$. This means that the x-values can be found for which $3x^2 - 8x + 4 = 25$ so providing the graphical solution of the quadratic equation $3x^2 - 8x - 21 = 0$.

(iii) Other tasks to be introduced in Chapter 13.

12.7 EXERCISES

1. Produce a table of values for the function $f(x) = 8x - 3$ for values of x from -4 to 5. Use it to plot the graph of the function $f(x)$.

2. Produce a table of values for the function $f(x) = 5 - 4x + 3x^2$ for values of x from -2 to 6. Use the table to plot the graph of the function $f(x)$.

3. Produce a table of values for the function $f(x) = x^3 - 7x^2 + 3x + 5$ for values of x from -3 to 7. Use the table to plot a graph of the function $f(x)$. Produce a second table of values for the function $g(x) = 3x + 4$ for values of x also from -3 to 7. Use the table to plot the graph of $g(x)$ on the same axes as used for the graph of $f(x)$.

   By inspection of any point(s) where the two graphs intersect, find the value(s) of x for which $f(x) = g(x)$.

12.8 POLYNOMIAL EXPRESSIONS

Different functions create different shaped graphs and it is worth investigating some of the more commonly encountered functions. A function $f(x)$ is that of a polynomial expression if it can be written in the form $f(x) = a + bx + cx^2 + dx^3 + ex^4 + fx^5 + gx^6 +$, etc where a, b, c, d, e, f, g, etc are all constant values. The different types of polynomial are the results of there being only a limited number of such terms and in the simplest case has $f(x) = a + bx$ (so that $c = d = e = f = g =$ etc $= 0$ from the more general polynomial expression).

The expression $f(x) = a + bx$ is known as a *linear function* since its graph will be seen to be a straight line. Such a function will always cross the y-axis (which measures $f(x)$) at the point $(0, b)$ and have a gradient of a. Hence the function $f(x) = 5x - 2$ will have a gradient of 5 and will cross the y-axis at $(0, -2)$. Gradient is defined to be the increase in the value of $f(x)$ for every 1 unit increase in the value of x; hence for $f(x) = 5x - 2$ there is $f(3) = 13$ and $f(4) = 18$ so that for a 1 unit increase in x, $f(x)$ has increased by 5 units, thus the gradient of the graph between these two values must be 5. It is characteristic of a linear function that its gradient is the same at all points. The graph of this particular function appears in the diagram shown in Figure 12.5.

The second form of polynomial function is the quadratic, in which $f(x) = ax^2 + bx + c$. The shape of the graph of the quadratic

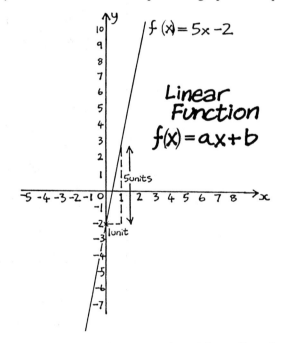

Figure 12.5   Polynomial Expression – Linear Function

function is that of a parabola but it has two alternative forms as seen in the next diagram, depending upon whether a is positive or negative. In the case in which a is positive then the parabola is 'nose-downwards' and it has a lowest point, or minimum value, when $x = -b/2a$. If a is negative then the parabola 'points' upwards and has a greatest or maximum value when $x = -b/2a$. In either case the graph cuts the y-axis at $(0,c)$ and it cuts the x-axis at the two points which represent the solutions of the quadratic equation $ax^2 + bx + c = 0$. It is possible for the curve to *just* touch the x-axis rather than cut it in two different points and this corresponds to the situation in which the equation $ax^2 + bx + c = 0$ has two equal solutions; if however, the graph fails to cut the x-axis at all, then this corresponds to the quadratic equation having no real solution at all. There is a close relationship between the quadratic function and its behaviour as seen from its graph in Figure 12.6 and the

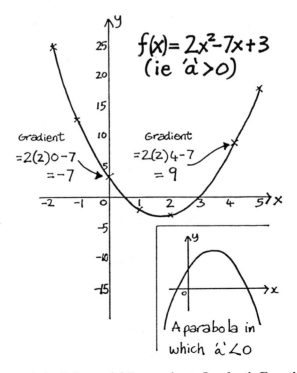

Figure 12.6   Polynomial Expression – Quadratic Function

solutions to a quadratic equation as described in Chapter 11.

Note that in the case of a quadratic function the gradient of the graph changes as x changes; it is, however, always equal to $2ax + b$ at any point $(x, f(x))$ on the curve. In the diagram showing the graph of $f(x) = 2x^2 - 7x + 3$ the gradient is positive from $x = 1.75$ upwards and the further to the right the steeper the gradient gets, whereas if x is less than 1.75 the gradient is negative since the shape is going 'downhill' from left to right.

The cubic function requires that $f(x) = ax^3 + bx^2 + cx + d$. Once again there are two characteristic shapes as seen in Figure 12.7, depending upon the sign of a the coefficient of x^3. In general such a curve will have one maximum and one minimum point, that is, one 'peak' and one 'trough'.

There are an infinite number of other possible polynomial functions and it is not possible to introduce them all; the linear, quadratic and cubic functions do provide an adequate introduction to the most commonly encountered polynomial functions. The shapes do, however, get increasingly involved as the number of terms in the polynomial expression increases.

One other function which should be encountered is that in

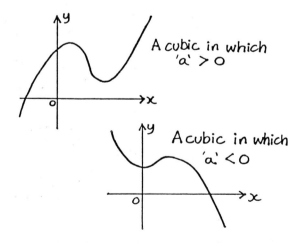

Figure 12.7   Polynomial Expression – Cubic Function

which $f(x) = \dfrac{1}{ax + b}$. The denominator is a linear function but the shape is very much *non*-linear and is seen in the diagram in Figure 12.8. Note that it is split into two parts which are quite separate and the function is *not defined* if $x = -b/a$, that is to say no value exists for $f(-b/a)$. Situations such as this may appear rather strange at first but they stem from the fact that if $x = -b/a$ then the denominator becomes zero and, no matter what the numerator, there is no value that can be ascribed to anything at all divided by zero; this accounts for the discontinuity in the curve when $x = 2.5$, in the example shown.

There is a wide variety of other similar functions in which the denominator is a polynomial as well as many in which *both* numerator *and* denominator are polynomials; their study is far more complex than that of the functions described so far in this chapter and will not be dealt with further in this book.

12.9   EXPONENTIAL FUNCTIONS

The phrase 'exponent' has been met previously in the context of the exponent form of floating-point representations in Chapter 5;

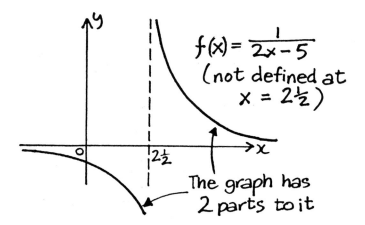

Figure 12.8   Polynomial Expression – Linear Function with Non-linear Expression

it refers to the power to which a number is raised. Hence the idea of an exponential function is one in which x is a power as in the case $f(x) = 2^x$. The diagram in Figure 12.9 shows three different exponential functions in which $f(x) = 2^x$, $g(x) = e^x$ and $h(x) = 3^x$; 'e' is an important mathematical quantity which has the value $2.7182818\ldots$ and, although not related to π at all, it does have the same sort of standing in mathematics as does π.

Note that all three curves exhibit the same characteristic shape and this is true for all functions of the type $f(x) = a^x$ provided that a is positive. They all pass through the point $(0,1)$ and are always completely above the x-axis, so that $f(x)$ can never be negative no matter what the value of x is.

12.10 CURVE SKETCHING

Plotting a curve requires a totally accurate drawing, whereas curve sketching requires only the creation of the essential shape of the

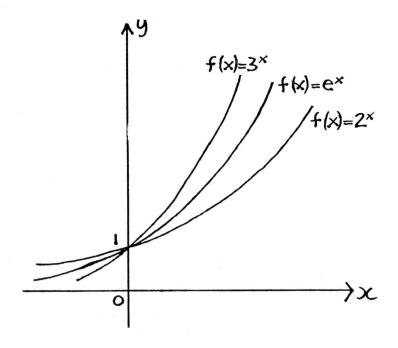

Figure 12.9   Exponential Functions

curve with very little attempt made to plot accurately. That does not mean to say that accuracy is deliberately abandoned but sketching is done for speed where accurate plotting is less important. In general, sketching will attempt to reproduce the outline shape of the curve and to identify certain key features such as the points where the curve crosses the x-axis and the y-axis or where it has 'peaks' or 'troughs'.

Obviously, if a computer can be used to produce an accurate plot of a curve as quickly or quicker than it can be sketched by hand then it must be preferred, but it is not always convenient to do this despite the accuracy of which computer plotting is capable.

12.10.1 Linear Functions

The general shape of a linear function is of course already known to be a straight line and it can be recognised by having a function of the form $f(x) = ax + b$ where a and b have constant values. In order to ensure a satisfactory sketch on axes which may have been drawn up quite quickly there are three important points to indicate:

(i)   where it crosses the x-axis (ie when $f(x) = 0$);

(ii) where it cross the y-axis (ie when $x = 0$);

(iii) that it *is* a straight line.

The gradient can also be checked that it is 'about right' so that if it is positive ($a > 0$ in $f(x) = ax + b$) then the line should slope up from *bottom left* to *top right*.

The diagram in Figure 12.10 shows the sketch of $g(x) = 2x + 3$ and the stages taken to sketch it.

12.10.2 Quadratic Functions

Having already seen that the general shape of a quadratic function is that of a parabola we are able to deduce at once from the sign of the coefficient of x^2 whether it is 'nose down' or 'pointing upwards' (by the coefficient being positive or negative respectively).

Hence the following checklist can be produced of points to note when attempting to sketch the curve of a quadratic function:

(i)   Find where it crosses the y-axis (ie when $x = 0$).

$$g(x) = 2x + 3$$

(i) Crosses x-axis if $2x + 3 = 0$
$$\therefore x = -1\tfrac{1}{2}$$

(ii) Crosses y-axis at $y = 3$

(iii) Gradient is 2 $\therefore$ positive

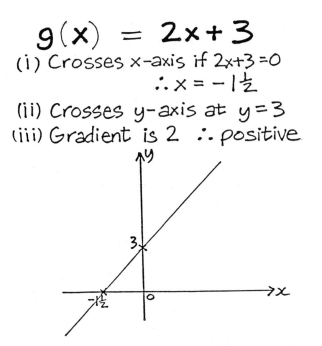

Note: No plotting of points except those given above, ie $(-1\tfrac{1}{2}, 0)$ and $(0, 3)$ and even those are not measured, just marked in approximately.

Figure 12.10   Curve Sketching – Linear Function

(ii)   Check the general shape by noting the sign of the coefficient of x^2.

(iii)   If possible find where it crosses the x-axis by solving the equation $f(x) = 0$. This is easy and very useful if $f(x)$ factorises but otherwise be careful not to take up too much time by the use of the 'formula' since, although it would be very useful to know just where the graph does cross the x-axis, the time taken to apply a formula solution may not be justified.

(iv)   For greatest accuracy try to find the maximum/minimum point. This always occurs when $2ax + b = 0$ or at $x = -b/2a$ so this value can be calculated; also find $f(-b/2a)$ to locate the point at which the 'peak' or the 'trough' occurs.

Figure 12.11 shows the use of these stages in the case of a sketch of the quadratic function $h(x) = x^2 - 3x - 10$.

12.10.3   Cubic Functions

The 'standard' forms of the cubic function were seen earlier in this chapter but there are relatively few key points to get hold of. However two can be identified:

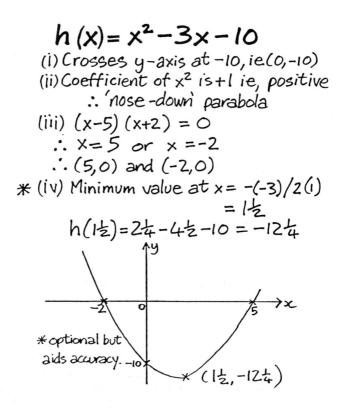

$$h(x) = x^2 - 3x - 10$$

(i) Crosses y-axis at -10, ie (0, -10)

(ii) Coefficient of x^2 is +1 ie, positive

∴ 'nose-down' parabola

(iii) $(x-5)(x+2) = 0$

∴ $x = 5$ or $x = -2$

∴ $(5, 0)$ and $(-2, 0)$

* (iv) Minimum value at $x = -(-3)/2(1)$

$= 1\frac{1}{2}$

$h(1\frac{1}{2}) = 2\frac{1}{4} - 4\frac{1}{2} - 10 = -12\frac{1}{4}$

*optional but aids accuracy.

$(1\frac{1}{2}, -12\frac{1}{4})$

Figure 12.11   Curve Sketching – Quadratic Function

(i)   Find where it crosses the y-axis (ie when $x = 0$).

(ii) Use the coefficient of x^3 to determine *which* of the standard shapes is likely.

Without the use of calculus treatment (which is beyond the scope of this book), it is difficult to find the positions of the maximum and the minimum points (Figure 12.12). It is still possible however, to be able to reinforce the idea of the 'standard' shape by reference to the function itself; if the function is

$$g(x) = 5x^3 - 6x^2 + 3x - 23$$

then the term which will have the greatest influence upon the value of $g(x)$ will be the $5x^3$ term in cases of large values of x. If x is large and positive it follows that so too is $5x^3$ and hence so is the whole of

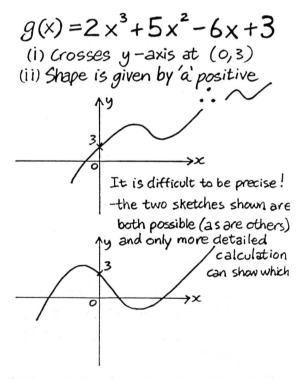

$g(x) = 2x^3 + 5x^2 - 6x + 3$
(i) Crosses y-axis at $(0, 3)$
(ii) Shape is given by 'a' positive

It is difficult to be precise! —the two sketches shown are both possible (as are others) and only more detailed calculation can show which

Figure 12.12   Curve Sketching – Cubic Function

$g(x)$; if x is large but negative then so too is $5x^3$ and hence so too is the whole of $g(x)$. This does not add to our ability to sketch the cubic function but it does reinforce the idea that the standard shapes are determined by the sign of the coefficient of x^3.

12.10.4 Exponential Functions

Since there is really only one standard shape, since it is bound to go through $(0,1)$ and since the whole of the graph is bound to be above the x-axis there is very limited scope for any variation here. The only differences that do exist relate to 'steepness' or gradient so that the graph of the function 5^x will go up more steeply for example than that of 4^x. This factor even so is only really of importance when attempting to show two exponential functions in the same sketch. The general shape of 5^x is shown in the diagram in Figure 12.13.

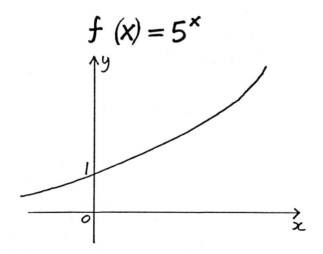

The only real distinction between different exponential curves is their slope NOT their characteristic shape

Figure 12.13   Curve Sketching – Exponential Function

12.11 EXERCISES

Sketch the graphs of each of the following functions:

(i) $f(x) = 3x + 4$;

(ii) $g(x) = -2x - 3$;

(iii) $f(x) = x^2 - 2x - 3$;

(iv) $g(x) = x^2 - 5x$;

(v) $f(x) = 6 + x - x^2$;

(vi) $h(x) = 4x^3 + 3x^2 - 7$;

(vii) $f(x) = 7^x$.

13 Using Graphs for Estimation Purposes

13.1 INTRODUCTION

As it is now possible for you to produce accurately plotted graphs of a variety of different functions an exploration of what can be done with them is now necessary. Some of the various possibilities have already been introduced, but there are four additional important areas to be discussed in this chapter. Interpolation and extrapolation are important ideas since they are involved in the concept and application of forecasting methods. Likewise the estimation of gradients and of areas under curves are important; these can be done by calculus methods but since this technique is beyond the scope of this book and also because its application sometimes is far too complex we need to investigate what can be done, and it is quite a lot, solely on the basis of accurately drawn graphs. Note that in this chapter there is no need for curve sketches except for the purpose of determining approximate shapes prior to accurate drawing.

13.2 INTERPOLATION AND EXTRAPOLATION INTRODUCED

If given a function which is defined by a formula on the lines of '$f(x) = \ldots$' then the value of $f(x)$ can be obtained for any value of x merely by substituting into the $f(x)$ equation. If however the function is defined by reference to a finite set of pairs of x and $f(x)$ values, then we do not *know* any $f(x)$ values for values of x other than those listed in the set of pairs. This will certainly be true if some experiment had been carried out leading to a series of observations being made at 1 minute intervals of, possibly, a temperature; the temperature may well be known after 7 minutes and also after 8 minutes but if asked subsequently to determine what the temperature was after 7¹/₂ minutes the value was not actually recorded. The same would apply to the recording of the trading profits of some company as a function of time or a patient's heartbeat recorded at 5 minute intervals; intermediate values cannot be determined without recourse to *interpolation* techniques.

Similarly if the trading profits for the years 1980 to 1987 inclusive have been recorded we may extend forwards, or *extrapolate,* to anticipate the figures for 1988; indeed it may be possible to go further and anticipate the 1989 figures. This is the concept of extrapolation and clearly it is an important one if one is to be able to determine trends for the future on the basis of evidence about the past or the present.

13.3   INTERPOLATION

Here, it is assumed that our knowledge extends simply to a set of discrete $(x, f(x))$ points on a graph and that we wish to estimate the value of $f(x)$ for some x-value intermediate between two points already plotted. Clearly if the formula '$f(x) = \ldots$' is available then substitution will give the required value; if using an automatic recording device plotting values on a continuous basis (as in the case of a heartbeat monitor or a temperature recording device which moves a pen over a rotating drum) then it is simply a matter of reading off the value(s) required.

For the first example in Table 13.1 it is assumed that we have two consecutive x-values and the corresponding $f(x)$ values, thus:

x	$f(x)$
3.6	24.63
3.7	28.76

Table 13.1   Data for Worked Example on Interpolation

and we wish to estimate $f(3.68)$. In this case and with no further information it has to be assumed that there is a linear relationship between x and $f(x)$ so that, since 3.68 is $^8/_{10}$ of the way between 3.6 and 3.7, we shall find that $f(3.68)$ is $^8/_{10}$ of the way between $f(3.6)$ and $f(3.7)$. Thus since the difference between the two $f(x)$ values is $28.76 - 24.63 = 4.13$ it follows that $^8/_{10}$ of this is 3.304. Thus $f(3.68) = 24.63 + 3.304$ or 27.93 to 2 dp as in Figure 13.1 (the same level of accuracy that was used previously for $f(x)$ values).

$$\text{Assuming a linear relationship find } f(3.68)$$
$$f(3.68) = f(3.6) + 0.8 \times \{f(3.7) - f(3.6)\}$$
$$= 24.63 + 0.8 \times 4.13$$
$$= 24.63 + 3.304 = \underline{27.934}$$

Figure 13.1   Example of Formula for Interpolation

However, if there are more than two pairs of values of x and $f(x)$ then it may be possible to establish a more sophisticated relationship and so calculate an intermediate value with greater accuracy. Suppose you are given Table 13.2:

x	$f(x)$
4.7	12.94
4.8	15.31
4.9	16.25

Table 13.2   Values for Calculating $f(4.87)$

and you wish to obtain $f(4.87)$ as accurately as possible. The method involves creating a table of differences as in Table 13.3:

		Differences	
		---	---
x	$f(x)$	1st, d_1	2nd, d_2
4.7	12.94	2.37	−1.43
4.8	15.31	0.94	−
4.9	16.25	−	−

Table 13.3   Table of Differences

in which d_1 for $x = 4.7$ is $f(4.8) - f(4.7)$ and d_1 for $x = 4.8$ is $f(4.9) - f(4.8)$. We cannot find d_1 for $x = 4.9$ since the value of $f(5.0)$ is not known. Then the second difference d_2 for $x = 4.7$ is d_1 for $x = 4.8 - d_1$ for $x = 4.7$. It is not possible to calculate any other differences with the values available here. Next is a formula to lead us to $f(4.87)$ based on the fact that 4.7 is the 'base' value and that 4.87 is 0.17 above that base value. Furthermore for the purposes of the formula x must be increasing in steps of 1 at a time and since in reality it has gone up in steps of 0.1, all x values (and the differences between them) will be scaled up by a factor of 10, thus the 0.17 referred to as being the amount of 4.87 above the base value will be treated as 1.7 in fact. In the formula the quantity just defined as having a value of 1.7 will be generalised as D. The formula is:

$$f(4.87) = f(4.7) + D(d_1 \text{ at } x = 4.7) + \tfrac{1}{2}D(D - 1) (d_2 \text{ at } x = 4.7)$$

$$= 12.94 + 1.7(2.37) + \tfrac{1}{2}(1.7)(0.7)(-1.43)$$

$$= 12.94 + 4.029 - 1.7017$$

$$= 15.2673 \text{ or } 15.27 \text{ to the same level of accuracy used}$$
$$\text{elsewhere for values of } f(x).$$

The formula introduced above has not been justified nor will any attempt be made to justify it; it is based upon more advanced mathematical ideas but it is important that it is *used* if one is to interpolate accurately. It uses a 'base' value of 4.7 since that is the lowest of the x-values in the original data and because both d_1 and d_2 can be worked out for this base value only. If we wanted instead to estimate the value of $f(4.74)$ then we have 0.04 above the base so that $D = 0.4$ (remembering the scaling factor) giving:

$$f(4.74) = f(4.7) + D(d_1 \text{ at } x = 4.7) + \tfrac{1}{2}D(D - 1)(d_2 \text{ at } x = 4.7)$$

$$= 12.94 + 0.4(2.37) + \tfrac{1}{2}(0.4)(-0.6)(-1.43)$$

$$= 12.94 + 0.948 + 0.1716$$

$$= 14.0596 \text{ or } 14.06 \text{ to 2 dp.}$$

If presented with four pairs of values of x and $f(x)$ rather than with three, then three difference values d_1, d_2 and d_3 will be required in addition to a revised formula. In the following example the x values are given as 6.3, 6.4, 6.5 and 6.6 with the corresponding $f(x)$ values and we wish to estimate the value of $f(6.43)$. The table also sets out the differences required for the formula, which is now:

$$f(6.43) = f(6.3) + D(d_1 \text{ at } x = 6.3) + \tfrac{1}{2}D(D - 1)(d_2 \text{ at } x = 6.3)$$
$$+ \tfrac{1}{6}D(D - 1)(D - 2)(d_3 \text{ at } x = 6.3).$$

The values are:

| x | $f(x)$ | Differences | | |
		1st, d_1	2nd, d_2	3rd, d_3
6.3	202.867	9.957	0.344	0.006
6.4	212.824	10.301	0.350	–
6.5	223.125	10.651	–	–
6.6	233.776	–	–	–

Table 13.4 Table of Differences to Obtain the Value of $f(6.43)$

Since $f(6.43)$ is to be estimated which is 0.13 above the base value and since we must scale up by 10, a D value of 1.3 will be adopted.

Thus we shall have:

$$f(6.43) = 202.867 + 1.3(9.957) + \tfrac{1}{2}(1.3)(0.3)(0.344)$$
$$+ \tfrac{1}{6}(1.3)(0.3)(-0.7)(0.006)$$

$$= 202.867 + 12.9441 + 0.06708 - 0.000273 = 215.878 \text{ to 3 dp.}$$

13.4   EXERCISES

1. Given that $f(4.7) = 12.96$ and $f(4.8) = 17.31$ estimate the value of $f(4.72)$.

2. Given that $g(6.9) = 71$ and $g(7.0) = 64.3$ estimate the value of $g(6.97)$.

3. Given that $f(3.2) = 215$, $f(3.3) = 227$ and $f(3.4) = 247$ estimate the value of $f(3.36)$.

4. Given that $f(3.16) = 49$, $f(3.17) = 51$ and $f(3.18) = 53$ estimate the value of $f(3.177)$ (NB Think carefully about scaling).

5. Given that $f(12.6) = 93.1$, $f(12.7) = 94.7$, $f(12.8) = 97.3$ and $f(12.9) = 100.2$, estimate the value of $f(12.76)$.

13.5   EXTRAPOLATION

Once again the concern is with finding values that were not included in the original set of values, but this time it is with values which lie *beyond* the range of those given. In the diagram shown in Figure 13.2, there are five points plotted P_1 to P_5. If we wanted to find the value of a point between P_1 and P_5 it would be possible to interpolate with some reasonable degree of accuracy, but if we wish to find $f(7.4)$ this is outside the given range and although the same techniques may be employed, involving the creation of a difference table and a revised formula, the more movement to the right of $x = 7$ or indeed to the left of $x = 3$, the less reliable the estimate becomes.

Since the use of this approach is needed for forecasting purposes, it is worth noting that more sophisticated methods do exist to determine the reliability of the results although these are beyond the scope of this book. In the example given here there are five pairs of x, $f(x)$ values so the formula becomes:

$f(7.4) = f(3) + D(d_1 \text{ at } x = 3) + \frac{1}{2} D(D-1)(d_2 \text{ at } x = 3)$
$+ \frac{1}{6} D(D-1)(D-2)(d_3 \text{ at } x = 3)$
$+ \frac{1}{24} D(D-1)(D-2)(D-3)(d_4 \text{ at } x = 3)$

The table of values is shown in Table 13.5:

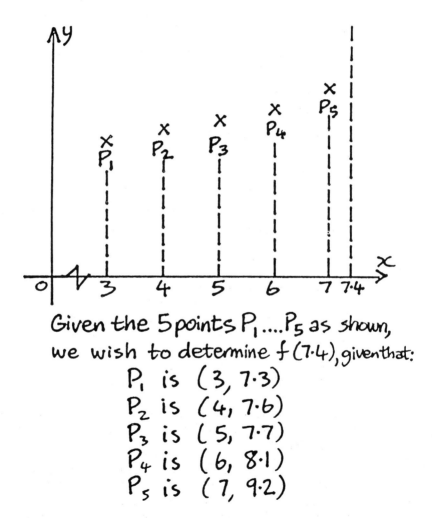

Given the 5 points P_1 .... P_5 as shown,
we wish to determine $f(7.4)$, given that:

P_1 is $(3, 7.3)$
P_2 is $(4, 7.6)$
P_3 is $(5, 7.7)$
P_4 is $(6, 8.1)$
P_5 is $(7, 9.2)$

Figure 13.2   Extrapolation Diagram

| x | $f(x)$ | Differences | | | |
		1st, d_1	2nd, d_2	3rd, d_3	4th, d_4
3	7.3	0.3	−0.2	0.5	−0.1
4	7.6	0.1	0.3	0.4	–
5	7.7	0.4	0.7	–	–
6	8.1	1.1	–	–	–
7	9.2	–	–	–	–

Table 13.5   Table of Values for Figure 13.2

Here, since x already goes up in steps of 1, no scaling is required so it is possible to regard D as 4.4. Thus:

$$f(7.4) = 7.3 + 4.4(0.3) + \tfrac{1}{2}(4.4)(3.4)(-0.2)$$
$$+ \tfrac{1}{6}(4.4)(3.4)(2.4)(0.5)$$
$$+ \tfrac{1}{24}(4.4)(3.4)(2.4)(1.4)(-0.1)$$

$$= 7.3 + 1.32 - 1.496 + 2.992 - 0.20944$$

$$= 9.90656 \text{ or } 9.9 \text{ to 1 dp.}$$

13.6   GRADIENTS

The concept of a gradient (or a slope) was introduced in the previous chapter. The next diagram shows the graph of a function f, and a point P with co-ordinates $(a, f(a))$ has been marked. We define the gradient of the graph at point P to be the rate at which $f(x)$ is changing per unit change in the value of x at that point. Whereas in the case of a straight line the slope is constant for the whole of its length, in the case of all other functions (such as the one shown here), it is changing all the time. In *this* case it gets steeper as it moves towards the right, so the gradient is getting larger.

Two methods exist for establishing the value of the gradient, the first (not dealt with in this book) being the more exact method using a calculus treatment. The second is an estimate, based upon the accuracy of the drawing (see Figure 13.3). The estimation method is quite sound as long as the gradient is not changing too rapidly in the vicinity of P. The tangent is drawn so as to 'touch' the curve at P as carefully as possible and it cuts the x-axis when $x = b$. It follows that the gradient is $f(a)$ divided by the length c ($= a - b$). Note in particular that the values of $f(a)$ and of c are not just the measured lengths on the drawing but what these lengths represent when the scales are taken into account. Hence if $f(a) = 6$ and $c = 12$ then the gradient must be $6/12$ or $1/2$ no matter what the lengths of the two lines may actually be in the diagram, so is totally independent of the scales used.

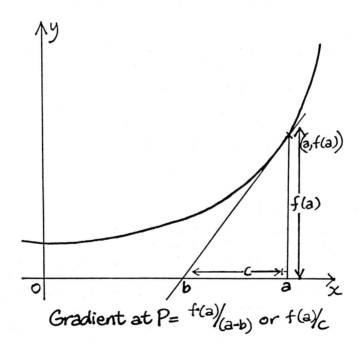

Figure 13.3   Example of Gradients

13.7   AREAS UNDER CURVES

It may occasionally be necessary to calculate the area under the graph of $f(x)$. If we are told that this is the area bounded by the curve, the x-axis and the *ordinates* at $x = 2$ and at $x = 10$ then it will be that which is shown shaded in the next diagram. The word 'ordinate' refers to a vertical line drawn from a point on the curve to the x-axis; so the ordinate at $x = 2$ in the diagram shown in Figure 13.4, is the line going vertically upwards from the point $x = 2$ on the x-axis until it meets the curve.

The area under the curve can be calculated exactly by a calculus

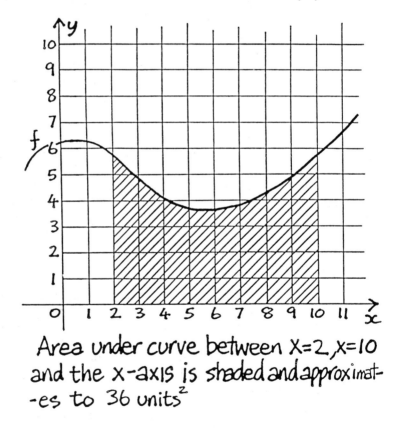

Area under curve between X=2, X=10 and the x-axis is shaded and approximates to 36 units²

Figure 13.4   Measuring the Area Under a Curve

method but only if we have the function defined by an equation of the type '$f(x) = \ldots$'. In other cases the value may be estimated by methods such as 'counting squares' as in the diagram or 'numerical integration'. Indeed it may frequently be found that we *have* to resort to estimating methods since it may not always be possible to apply more exact methods.

If the graph has been plotted accurately the squares shown in the shaded area can be counted up, usually employing some convention whereby those squares which are part in and out of the shaded area are counted in if they are more in than out and vice versa. The area of each square is not just 'one'; it depends upon the scales on the two axes and what these axes represent. If, for example, the x-axis measures time in seconds and the y-axis is to represent speed in metres per second then the area measures time × speed, or distance, in metres. Particular care must be taken if the time-axis is in minutes and the speed is in metres per second for example, since then the squares are in units of 60 metres each, but still of course representing distance.

There are also two methods which are classified as being 'numerical integration' methods known as the Trapezium Rule and Simpson's Rule and are dealt with in the next two sections. They rely more on the tabulated values of x and $f(x)$ instead of the rather more crude square counting method; although still representing an approximate method they can prove to be capable of producing as refined a level of accuracy as may ever be required.

13.8 THE TRAPEZIUM RULE

If an estimate of the area under the curve is required between the x-axis and the ordinates at $x = 2$ and at $x = 18$, this can be done by subdividing the area into an *even* number of vertical strips, each of width w. If we use y_0 and y_n as the heights of the first and the last ordinates then it is possible to add together the areas of all the strips, (each of which is a trapezium in shape), to yield the formula quoted in Figure 13.5.

In Figure 13.5 the strips are accorded dotted boundaries and it is easy to see that whilst areas such as ABCD produce an under-estimate of the true area under the curve, these are compensated

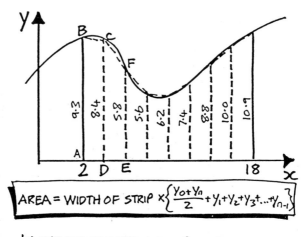

IN THE EXAMPLE THERE ARE 8 STRIPS COVERING

$X=2$ UP TO $X=18$ ∴ THE WIDTH OF EACH IS

$\frac{18-2}{8} = 2$ $Y_0 =$ HEIGHT OF 1ST COLUMN $= 9.3$

$Y_n =$ HEIGHT OF LAST COLUMN $= 10.9$

AREA $= 2 \times \left\{ \frac{9.3+10.9}{2} + 8.4 + 5.8 + 5.6 + 6.2 + 7.4 + 8.8 + 10.0 \right\}$

$\qquad = 2 \times \left\{ 10.1 + 52.2 \right\} = 2 \times 62.3$

$\qquad\qquad = 124.6$ SQUARE UNITS

Figure 13.5   The Trapezium Rule

for by areas such as the one with ordinates of 5.8 and 5.6 which gives an overestimate of the true area. If we were to increase the number of strips then the width w is reduced, so bringing the trapezia closer and closer to the 'true' area under the curve, allowing such a process to yield as accurate a result as may be desired, by choice of a sufficiently large number of strips. The heights of the ordinates may be determined by the '$f(x) = \ldots$' formula or derived from values observed, for example, in an experiment. Note, however, the necessity for there to be an *even* number of strips.

13.9   SIMPSON'S RULE

This is a rather more sophisticated method than the Trapezium

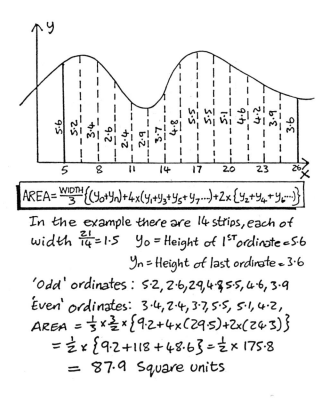

In the example there are 14 strips, each of width $\frac{21}{14} = 1.5$ $y_0 =$ Height of 1st ordinate $= 5.6$

$y_n =$ Height of last ordinate $= 3.6$

'Odd' ordinates: $5.2, 2.6, 2.9, 4.8 \cdot 5.5, 4.6, 3.9$

'Even' ordinates: $3.4, 2.4, 3.7, 5.5, 5.1, 4.2,$

$AREA = \frac{1}{3} \times \frac{3}{2} \times \{9.2 + 4 \times (29.5) + 2 \times (24.3)\}$

$= \frac{1}{2} \times \{9.2 + 118 + 48.6\} = \frac{1}{2} \times 175.8$

$= 87.9$ Square units

Figure 13.6 Simpson's Rule

Rule but again requires the subdivision of the area into an even number of strips. This time, however, whilst it still requires a knowledge of the heights of the first and the last ordinates y_0 and y_n respectively, it also subdivides the rest of the ordinates into 'odd' ones such as $y_1, y_3, y_5,$ etc as well as the 'even' ones such as $y_2, y_4, y_6,$ etc.

This is no more difficult a method to use than the Trapezium Rule and, like it, it is capable of a high degree of accuracy by increasing the number of strips. Using the same number of strips it is slightly more accurate a method than is the Trapezium Rule. Do note however, the absolute necessity for referring to the first ordinate as y_0 and *not* as y_1. That apart, its use is quite straightforward. In Figure 13.6 Simpson's Rule is used to estimate the area under the curve as shown; it has been drawn with 14 strips and covers the area contained between the curve, the x-axis and the ordinates at $x = 5$ and at $x = 26$.

13.10 EXERCISES

1. Draw the graph of $f(x) = x^2 - 4x + 7$ as accurately as possible between $x = -3$ and $x = 5$. By drawing tangents at the points on the graph for which $x = -1$, $x = 2$ and $x = 4$, estimate the gradient of the curve at each of these three points.

2. By drawing the graph $f(x) = 3x^2 - 7x + 9$ as accurately as possible between the values of $x = -2$ and $x = 4$ estimate the area between the curve, the x-axis and the ordinates at $x = -2$ and $x = 4$ using the 'counting squares' method.

3. A function $f(x)$ is defined for a range of values of x by the following table:

x	3	4	5	6	7	8	9	10	11
$f(x)$	13	11	9	7	7	8	12	14	18

By using the Trapezium Rule with eight strips estimate the area between the curve, the x-axis and the ordinates at $x = 3$ and $x = 11$.

4. Repeat question 3 but this time using Simpson's Rule to estimate the same area.

14 Sets and Venn Diagrams

14.1 INTRODUCTION – WHAT IS A SET?

In its simplest form a set is no more than a collection of objects; the objects may be numbers, they may be the names of people, they may be colours or they may be a list of all the products of a software house. Perhaps the most important factor about a set is that there *must* be some *common feature,* which links together all the objects in some way; for example, perhaps the numbers may all be divisible by 5, the people may all work in the same department, the colours may be those seen in a rainbow and so on.

Any one object, or item, within a set is referred to as an *element* of the set, and we may identify specific sets either by giving them names or, more conveniently, by denoting them by capital letters. Hence A may be used to denote the set of numbers divisible by 3 between 1 and 25 inclusive, and we may list its elements as:

$$A = \{ 3, 6, 9, 12, 15, 18, 21, 24 \}$$

Thus there are two ways of describing set A, the first by defining it in words and the second by listing its elements. Frequently it will be found that the second method may prove more useful than the first for purposes of processing the set. It should be noted however, that if the elements *are* listed, as above, the order in which they are listed has absolutely no significance, so it could just as easily have been defined as:

$$A = \{ 9, 18, 24, 3, 6, 15, 21, 12 \}$$

Next, a notation is introduced which identifies that a particular quantity is an element of set A. To indicate that 3 is an element of A we could write:

$$6 \in A$$

and, likewise to indicate that 5 is *not* an element of set A, we write:

$$5 \notin A$$

Thus if M is used to refer to the set of computer manufacturers, we could say that Honeywell Bull $\in$ M and that Goodyear Tyre and Rubber Company $\notin$ M.

Another commonly used aspect of the terminology is to refer to *subsets*. Thus set B is a subset of set A if *every* element of B is also included within A. Hence if A is defined to be the set of all people who work for a particular company, and B to be the set of all people who work within the Data Processing Department of that same company, it follows that B is a subset of A; this is stated using symbols, as B $\subset$ A. The same approach as before is used if we wish to say that D is *not* a subset of A, which is written D $\not\subset$ A.

The last fundamental feature to be introduced is that if an element occurs *more* than once in a set it is *listed* only once. In a numerical example the factors of 120 (which are 2 x 2 x 2 x 3 x 5) may be referred to as *F*, defined as { 2, 3, 5 }. As an alternative example, we might find that in a data processing department, the programmers skilled in the use of an assembler language are Stephen, Lakbir, Rachel and Anna, whereas those who are very competent in Cobol are Lakbir, Anna, Sean and Amarjit; if we had to list those who were good with assembler language or Cobol, the set of such people P is given by:

$$P = \{ \text{Stephen, Lakbir, Rachel, Anna, Sean, Amarjit} \}$$

noting once more that the order is *not* important.

Set notation is a very useful means of classification. It is of great importance in the study of logic and this is the prime reason for its introduction here. Logic, as far as the computer is concerned, is not just a philosophy; it is directly involved with the physical make-up of the machine, as well as the way in which programming statements are executed and its study is therefore totally justified on either of these two counts.

14.2 THE UNIVERSAL SET

In any one situation, no matter what sets are used and what elements exist within these sets, there will be one set called the *Universal Set* and represented by $\mathscr{E}$, which contains every element in each of the other sets, so that every set used in a problem is a subset of the Universal Set. Unfortunately, there is likely to be a different Universal Set for each situation despite the title; in other words, it is universal only for a particular situation. The Universal Set therefore determines the 'boundaries' imposed by the situation, but sometimes it may be found that the Universal Set given provides wider boundaries than are needed within the problem.

Consider the situation of the personnel employed within the Data Processing Department of NatComp Co. We may list as set W all the female employees of that department, as M all the male employees, as O all the operators, as S all the analysts (male *or* female) and so on. Thus many different sets, M, W, O, S, etc can be defined but since each of them *must* be a subset of the list of all the employees working within the Data Processing Department that same list may conveniently constitute the Universal Set. Hence $\mathscr{E}$ may be classified to be the set of all employees working in the Data Processing Department of NatComp Co. It is possible to consider using the set of *all* employees of NatComp Co as $\mathscr{E}$, whether or not they work in data processing; this would still be quite feasible but would provide far wider boundaries than may be needed in the problem. Clearly however, if the range of the situation was extended so as to involve staff in other departments, then a more appropriate Universal Set would need to be found in the process.

14.3 EXERCISES

Given that $\mathscr{E}$ = { Red, Blue, Black, Green, Yellow, Pink, Orange }, that the set A = { Red, Blue, Green }, B = { Yellow, Pink, Blue } and that set C = { Black, Blue, Red, Green, Yellow }, decide whether the following statements are true or false:

(i)   Blue $\in$ A

(ii)   A $\subset$ C

(iii)   B $\not\subset$ A

(iv)   Green $\notin$ B

(v)   Red ∈ B   (ix)   { Blue, Yellow } ⊄ A

(vi)   C ⊂ A   (x)   Yellow ∉ A

(vii) { Red, Green } ⊂ A   (xi)   Yellow ∈ C

(viii) { Green, Red } ⊂ B   (xii) Orange ∈ C

14.4 UNION AND INTERSECTION

Once two sets A and B have been defined then we can define the *union* of A with B (written as A ∪ B) to be the set of elements which occur either in set A or in set B or in both, although elements which do occur in both are not listed more than once in the union. Hence, if A = { a,b,c,d } and B = { c,d,e,f } then A ∪ B = { a, b, c, d, e, f }. Similarly, if P = { 3, 6, 9, 12, 15 } and Q = { 6, 12, 18, 24 } then we can describe P ∪ Q = { 3, 6, 9, 12, 15, 18, 24 }.

This concept can easily be extended to the union of three or more sets, so that, if P = { Cobol, RPG11, Algol }, Q = { Algol, Fortran, Pascal, Ada } and R = { Basic, Algol, Cobol, Pascal } then:

P ∪ Q ∪ R = { Cobol, RPG11, Algol, Fortran, Pascal, Ada, Basic }.

Other examples are in Figure 14.1.

$$eg\ A = \{ 3, 6, 9, 12, 15, 18 \}$$
$$B = \{ 5, 10, 15, 20 \}$$

$$A \cup B = \{ 3, 5, 6, 9, 10, 12, 15, 18, 20 \}$$

$$eg\ A = \{ e, x, a, m \}$$
$$B = \{ t, r, i, a, l \}$$
$$C = \{ s, t, r, a, i, n \}$$

$$A \cup B \cup C = \{ a, e, i, l, m, n, r, s, t, x \}$$

Note that the order of the elements in the union is of no significance

Figure 14.1   Examples of Union

The other term to be introduced is the *intersection* of two sets A and B (written as A ∩ B) which defines those elements which exist *both* in set A *and also* in set B. Thus if

A = { 2, 5, 8, 11, 14, 17, 20 }, B = { 5, 10, 15, 20, 25, 30 }

then the intersection can be defined A ∩ B = { 5, 20 }. This may of course be extended to the intersection of three or more sets so that A ∩ B ∩ C is the set of elements which can be found *both* in A *and also* in B *and also* in C (Figure 14.2).

$$eg \ A = \{3, 6, 9, 12, 15, 18\}$$

$$B = \{5, 10, 15, 20\}$$

$$A \cap B = \{15\}$$

$$eg \ A = \{e, x, a, m\}$$

$$B = \{t, r, i, a, l\}$$

$$C = \{s, t, r, a, i, n\}$$

$$A \cap B \cap C = \{a\}$$

Figure 14.2   Examples of Intersection

14.5   EMPTY SETS

A set which contains no elements at all is called an *empty set* or sometimes a *null set*. It is denoted either by { } or by ∅. Empty sets occur in circumstances such as defining the intersection of two sets which have no elements in common; hence if A = { 7, 14, 21, 28 } and B = { 5, 10, 15, 20, 25, 30 } then A ∩ B = { }. Similarly if we were to refer to the set of programmers able to write a 1000-line Cobol program in half-an-hour this might very well be expected to be an empty set.

14.6 COMPLEMENTS

Given that it is possible to define a set A then we can also define A', the set of elements which *exist in $\mathcal{E}$ but which are not in* A. The set A' is called the complement of set A. Thus if $\mathcal{E}$ = { all integers between 1 and 15 inclusive } and if A = { all even numbers } then A = { 2, 4, 6, 8, 10, 12, 14 } and so A' = { 1, 3, 5, 7, 9, 11, 13, 15 } or this could be stated as A' = { odd numbers }. It is necessary to remember that A does *not* include 16, 18, 20, etc since these are excluded by virtue of the definition of $\mathcal{E}$.

It follows at once that A' *together with* A comprises the Universal Set and also that A $\cap$ A' = { }. There is no reason why the idea of a complement cannot be extended to a set such as A $\cap$ B; hence we could define (A $\cap$ B)'. It is interesting to note that this is the same as A' $\cup$ B', but no proof is offered here. Equally well (A $\cup$ B)' = A' $\cap$ B'; both of these relationships will become easier to appreciate with the introduction of Venn diagrams later in this chapter.

In the diagram which follows, note especially the slightly

$$\mathcal{E} = \{ x : 1 \le x \le 20, x \text{ integral} \}$$
to define the Universal Set.

$$A = \{ x : x \text{ is a multiple of } 3 \}$$
$$B = \{ x : x \text{ is an even number} \}$$

$$\therefore A' = \{ 1, 2, 4, 5, 7, 8, 10, 11, 13, 14, 16, 17, 19, 20 \}$$

$$B' = \{ 1, 3, 5, 7, 9, 11, 13, 15, 17, 19 \}$$
or
$$B' = \{ x : x \text{ is an odd number} \}$$

$$A \cup B = \{ 2, 3, 4, 6, 8, 9, 10, 12, 14, 15, 16, 18, 20 \}$$

$$\therefore (A \cup B)' = \{ 1, 5, 7, 11, 13, 17, 19 \}$$

Figure 14.3   Universal Set and Complements

different way of describing a set. A = { x:x is a multiple of 3 } reads as 'A is the set of x-values such that each x is a multiple of 3' or we could express this as A = { multiples of 3 }. It may appear to be rather more complicated to express set definitions in this way but it can have some advantages, especially in cases where the textual description is wordy and cumbersome; in such cases it is often possible to shorten the definition by the use of the ' { x:x . . .} etc' approach. Compare for example the Universal Set definition given in Figure 14.3 and the alternative:

$\mathscr{E}$ = { whole numbers which lie between 1 and 20 inclusive }.

Using this example, and referring back to the earlier statement that (A ∪ B)' = A' ∩ B' consider both A' and B' and list A' ∩ B'; the result will be found to be given by { 1, 5, 7, 11, 13, 17, 19 }. It is not a coincidence that this is the same as the stated result for (A ∪ B)'.

14.7   EXERCISES

1.   Defining $\mathscr{E}$ as { whole numbers between 3 and 19 inclusive }, A = { multiples of 3 } and B = { multiples of 7 } list the sets:

   (i)   A   (v)   A'

   (ii)   B   (vi)   B'

   (iii) A ∪ B   (vii) (A ∪ B)'

   (iv) A ∩ B   (viii) (A ∩ B)'

2.   Using the definition that $\mathscr{E}$ = { x: 5 ≤ x ≤ 26, x integral }, and with:

      A = { x: x is a multiple of 5 }

      B = { x: x is a prime number }

      C = { 9, 16, 25 }

   (i)   Describe C in words.

List each of the following:

   (ii)   A   (iv) A ∩ C

   (iii) B   (v)   A ∩ B

(vi)	B ∩ C		(xii)	A'
(vii)	A ∪ B		(xiii)	B'
(viii)	(A ∩ C)'		(xiv)	C'
(ix)	(A ∪ B)'		(xv)	A' ∩ B'
(x)	(B ∪ C)'		(xvi)	B' ∪ C'
(xi)	(B ∩ C)'		(xvii)	A' ∩ B' ∩ C'

14.8   VENN DIAGRAMS TO REPRESENT SETS

A *Venn diagram* is nothing more than a pictorial representation of sets and their interrelationships. They are a great deal more convenient than giving textual descriptions and they can sometimes be more convenient than listing elements. The following diagram Figure 14.4 shows how A ∪ B can be represented.

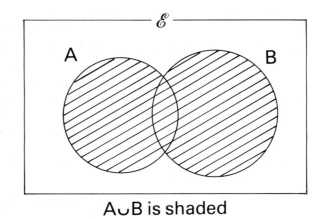

A∪B is shaded

Figure 14.4   Venn Diagram Representing A ∪ B

The elements that are contained in the various sets can be listed as shown in Figure 14.5.

From the diagram it can be immediately identified that
$$A = \{ 3, 6, 9, 12, 15, 18 \}$$
and that $B = \{ 2, 4, 6, 8, 10, 12, 14, 16, 18, 20 \}.$

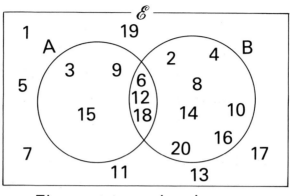

Elements may be shown

Figure 14.5   Venn Diagram Showing Elements of a Set

The intersection A ∩ B can be seen to yield the set { 6, 12, 18 }. We can see (A ∪ B)' = { 1, 5, 7, 11, 13, 17, 19 } and so on.

Note however, that the positions of the elements in the diagram is critical only in so far as they are in the correct region of the Venn diagram; the 6, 12 and 18 in A ∩ B could have been in any order just as if we had listed A ∩ B. Also, the amount of physical space given over to each set in the diagram has absolutely no significance; the fact that one set appears larger is meaningless since the number of elements that a set contains cannot be determined whatsoever from the space given over in the diagram to the representation of that set.

Although referred to in the previous diagram the intersection of the two sets P and Q can be seen clearly shaded in the next Venn diagram in Figure 14.6.

Sometimes we may have two sets such as A and B for which A ∩ B = { }. Such sets may be referred to as being *disjoint* as shown in Figure 14.7.

Note how, in each of the given diagrams, the Universal Set can be clearly seen to contain within it the sets we are dealing with and to provide a 'boundary' within which the situation may be described.

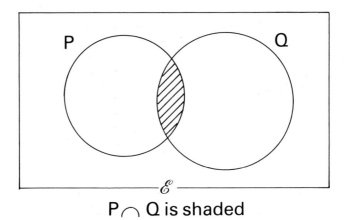

P∩ Q is shaded

Figure 14.6   Intersection of Two Sets

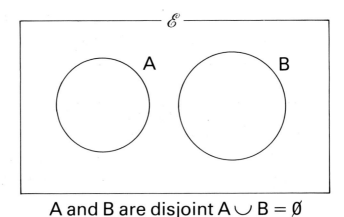

A and B are disjoint A ∪ B = Ø

Figure 14.7   Disjoint Sets

The next three diagrams (given in Figure 14.8) show how complements may be indicated in a Venn diagram in the case of two sets A and B; in the first A′ is shaded leaving A unshaded. In the second (A ∪ B)′ is shaded and in the third (A ∩ B)′. In each of these cases the elements have not been included in order to show more clearly what is meant by the complement in each case.

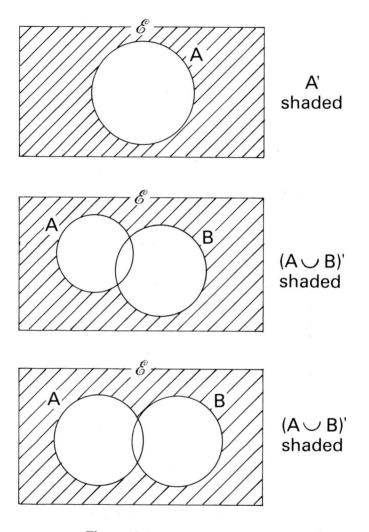

Figure 14.8   Complement of Two Sets

The final diagram in this section (Figure 14.9) illustrates the case of three sets A, B and C. It shows all the unions and intersections clearly and it lists a number of these (but not all those possible).

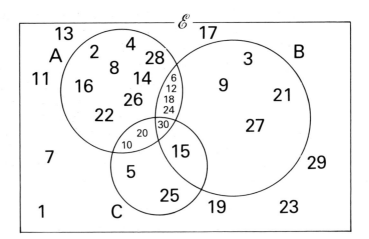

$$\mathcal{E} = \{x : 1 \leqslant x \leqslant 30\}$$
$$A = \{x : x \text{ is even}\}$$
$$B = \{x : x \text{ is divisible by } 3\}$$
$$C = \{x : x \text{ is divisible by } 5\}$$
$$\therefore A \cap B = \{10, 20, 30\}$$
$$B \cup C = \{3, 5, 6, 9, 10, 12, 15, 18, 20, 21, 24, 25, 27, 30\}$$
$$(A \cup B \cup C)' = \{1, 7, 11, 13, 17, 19, 23, 29\}$$
$$(B \cap C)' = \{1, 2, 3, 4, 5, 6, 7, 8, 9, 10, 11, 12, 13, 14, 16, 17, 18, 19, 20, 21, 22, 23, 24, 25, 26, 27, 28, 29\}$$
$$A \cap B \cap C = \{30\}$$

Figure 14.9   Union and Intersection of Three Sets

14.9   EXERCISES

1. Given that $\mathcal{E} = \{x : 0 < x < 17, x \text{ integral}\}$, $P = \{4, 8, 12, 16\}$ and $Q = \{3, 6, 9, 12, 15\}$ draw up a Venn diagram to illustrate

the sets $\mathcal{E}$, P and Q and their interrelationships. Mark all elements in the correct regions of the diagram. *From the diagram* list P $\cap$ Q and (P $\cup$ Q)'.

2. Given that $\mathcal{E}$ = { Adam, Beryl, Chandrakant, Desmond, Elaine, Fearon, Guy, Henrietta, Idris, Jacquetta, Kathryn, Leo }, being the set of data processing staff, the set of Cobol programmers C = { Beryl, Desmond, Idris, Kathryn, Leo } and the set of Fortran programmers F = { Adam, Beryl, Desmond, Guy, Jacquetta, Leo }. Represent the foregoing information by means of a fully annotated Venn diagram and use this to answer the following questions:

   (i)   How many people can program in Fortran?

   (ii) Who can program *both* in Cobol *and* in Fortran?

   (iii) Who can program in *neither* Cobol *nor* Fortran?

   (iv) How many people can program in Cobol but not in Fortran? The set which represents this is C $\cap$ F'.

   (v)   How many people can program in Fortran but not in Cobol?

   (vi) List those who can program either in Cobol or in Fortran or in both.

3. Given $\mathcal{E}$ = { x: 12 < x < 29, x integral }, P = { 14, 21, 28 }, Q = { 18, 24 } and R = { 16, 20, 24, 28 }, draw a Venn diagram to represent the given sets and their interrelationships, marking all elements in the correct regions. On the basis of the diagram list the following:

   (i)   P $\cup$ Q   (v)   (P $\cup$ Q $\cup$ R)'

   (ii) P $\cap$ Q   (vi) P $\cap$ R

   (iii) P $\cap$ Q $\cap$ R   (vii) Q $\cap$ R

   (iv) (P $\cap$ Q)'   (viii) (P $\cup$ R)'

14.10 VENN DIAGRAMS AND LOGICAL OPERATIONS

The concepts of logic are embedded in what is called *Boolean algebra* (named after George Boole who pioneered it). Any logical 'proposition' consists of a statement which is capable only of being either *true* or *false* and never anything else; thus 'you are sitting down' is such a statement and either it is true or it is false, there being no possible third category. In the same way the statement 'the sky is blue' is also a logical proposition. For convenience any logical proposition will be denoted by a single letter such as p or q and we shall refer also to the negation of the proposition by −p or ~p; thus if p is the proposition 'It is snowing' then −p is the proposition 'It is not snowing'. Clearly if p is true then −p must be false and vice versa.

The concepts of Boolean algebra can be extended to the study of gates, which are the basic building blocks of logic circuits. They are devices containing one or more inputs and producing a single output. By the use of these gates can be created the half adders, the full adders, flip-flops and other units which enable calculations to be performed in the Arithmetic and Logic Unit (ALU) or control to be exercised in the Central Processing Unit (CPU). They can be used to convert input decimal values into BCD values, to control hardware and to allow for interfacing between the CPU and its peripherals.

It is possible to represent propositions, their negations and other related functions by the use of Venn diagrams in the following manner:

(a) In Figure 14.10 the proposition −p is shown shaded whilst p is unshaded thus letting p and −p have equivalence with the set A and its complement A'.

(b) Here are two propositions p and q. The compound proposition p OR q can be taken as meaning that '*either* p is true *or* q is true *or* both are true'. This compound proposition is commonly

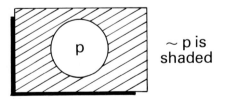

Figure 14.10   Venn Diagram Representing a Proposition

encountered in programming when implementing conditional statements such as the Cobol statement:

IF VAL-A < 25 OR GROSS NUMERIC PERFORM A-PARA VARYING X FROM 1 BY 2 UNTIL X = 25.

The Venn diagram approach makes this equivalent to the union of the sets A and B so that the shaded area is p OR q (or A ∪ B) as in Figure 14.11.

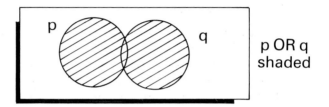

Figure 14.11   Venn Diagram Representing a Compound Proposition p OR q

(c)   This time we explore the compound proposition p AND q which can be taken as meaning '*both* p is true *and also* q is true'. Again this is common in programming statements which require both of two simple conditions to be true at one and the same time. In the Venn diagram given in Figure 14.12 the shaded area is p AND q, which is equivalent therefore to the intersection of two sets A ∩ B.

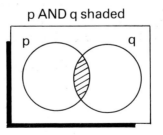

Figure 14.12   Venn Diagram Representing a Compound Proposition p AND q

The similarity of Venn diagrams and therefore of the inter-relationships between sets with logical propositions will be further exploited in more detailed work on logic in Chapters 15 and 16.

15 Boolean Algebra – The Algebra of Logic

15.1 LOGICAL OPERATIONS

15.1.1 NOT

In the previous chapter an introduction was given to the concept of a proposition and its negation. We may refer to a proposition as p and to its negation as −p or as NOT p. Quite clearly we shall need to be able to analyse any logic circuit in such a way that, if given the inputs to it, we can determine what the output will be; in the case of the very simple NOT-circuit (or inverter) there is just one input and one output. The input is the 'value' of the proposition p and this is an indication as to whether it is true (T) or false (F); since a proposition *must* be either true or false the value of the proposition can be defined to be T or F in every case. It is possible to therefore list all the different values of p and so tabulate the equivalent values of NOT p as shown in the following truth table (Figure 15.1).

Note that two slightly different forms have been given. Since T and F are a two-state system they can immediately be likened to the binary values 1 and 0. Furthermore since we may want to store

Figure 15.1 'Not' Truth Table

237

truth values in the computer it makes sense to maintain this relationship so the value of p will be defined as 0 if it is false and as 1 if it is true. The truth table is headed by the input proposition and the output; these are p and NOT p respectively (or ~ p). Below the input heading are listed all the values it can take (T and F or 1 and 0) and the entries under the output are the corresponding values for NOT p.

In a logic circuit there is also a need to represent a NOT-gate and this is shown below in Figure 15.2; the effect of such a gate is to invert an input signal (the 0 or the 1 value of p) to create the output signal (1 or 0 respectively).

NOT-GATE

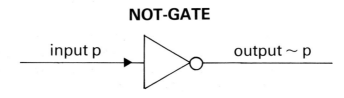

Figure 15.2   Not-gate for a Logic Circuit

The gate therefore inverts the input from a 'low' signal (0) to a 'high' signal (1) or vice versa.

15.1.2   OR

This compound proposition was also introduced in the previous chapter (see 14.4) and was defined there. Here the relevant truth table for this proposition is introduced in Figure 15.3, based upon the various combinations of the values of the two inputs p and q. The only way that this compound proposition can be false is if *both* input values are false and in no other way.

It is worth taking note that the word OR is used here in its 'inclusive' sense and that a different logical operation does also exist in which the 'exclusive OR' is defined as meaning that a compound proposition is true if '*either* p is true *or* if q is true *but not if both* are true'; at this stage there is no need to worry about the exclusive OR, other than to be warned that it does exist. Alternative names exist for the inclusive OR, describing it as the 'union' of the

OR

p	q	p OR q
0	0	0
0	1	1
1	0	1
1	1	1

TRUTH TABLE

Figure 15.3 'Or' Truth Table

two propositions (since it is equivalent to the union of two sets, as seen in the Venn diagram of the previous chapter), sometimes as the 'logical sum' (hence we shall see + used later on) or as the 'disjunction' of the two propositions.

The next diagram in Figure 15.4 shows the method of representing OR in a circuit, noting the symbol in use:

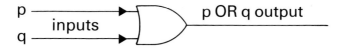

Figure 15.4 'Or' Represented in a Circuit

It is possible however to have three or more inputs to an OR-gate. In the diagram in Figure 15.5 the circuit representation of such a situation is shown with the truth table which defines its effect. Once again *the output is false only if all the inputs are false.*

Take particular note of the way in which the various combinations of values of p, q and r were shown in the table, to ensure that all combinations were listed without duplication or exclusion. The third column, for r, shows alternate 0 and 1, the preceding column, for q, shows alternately two 0s and two 1s, whilst the column for p consists alternately of four 0s and four 1s; had there been four

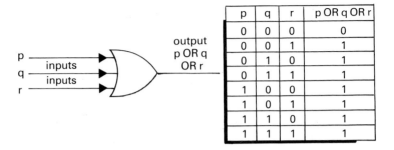

p	q	r	p OR q OR r
0	0	0	0
0	0	1	1
0	1	0	1
0	1	1	1
1	0	0	1
1	0	1	1
1	1	0	1
1	1	1	1

Figure 15.5   'Or' Circuit with Truth Table

inputs then the next column to the left would have had eight 0s and
eight 1s and so on.

15.1.3   AND

This compound proposition is defined by the truth table in Figure
15.6; there is an immediate contrast with the OR truth table since
this time the compound proposition can be true only if both input
propositions are true and is false in all other cases:

TRUTH
TABLE

p	q	p AND q
0	0	0
0	1	0
1	0	0
1	1	1

Figure 15.6   'And' Truth Table

Alternative names for this compound proposition are 'conjunction' or 'intersection' (compare this with the similarity with the
intersection of two sets) or 'logical product' (which will recur
later when a multiplication sign is introduced, although it will
be . rather than ×).

The AND-gate also appears in a logic circuit, in which case it appears in Figure 15.7.

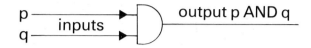

Figure 15.7   And-gate in a Logic Circuit

Equally well there can be three or more inputs into an AND-gate for which an example of the circuit and the truth table appear in Figure 15.8. Note that the output is true *only if all inputs are true* and in no other case.

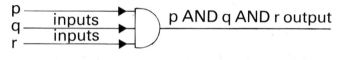

p	q	r	p AND q AND r
0	0	0	0
0	0	1	0
0	1	0	0
0	1	1	0
1	0	0	0
1	0	1	0
1	1	0	0
1	1	1	1

Figure 15.8   'And' Circuit with Truth Table

15.2   AN INTRODUCTION TO COMBINING LOGICAL OPERATIONS

Although in the early part of this chapter the three basic logic gates were encountered they were in each case the only gate in the circuit. In practice it is likely that circuits which contain a mixture of such gates will be developed, as for example in the case shown in Figure 15.9.

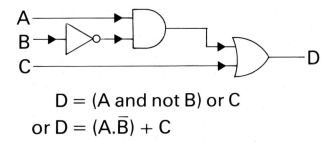

$$D = (A \text{ and not } B) \text{ or } C$$
$$\text{or } D = (A.\overline{B}) + C$$

Figure 15.9   Logic Circuit with Three Logic Gates

In this circuit B is immediately input into a NOT-gate so emerging as NOT-B (or $\overline{B}$, an easier way of writing $-B$ when using the algebraic form of the circuit). Then A and $\overline{B}$ are input into an AND-gate to produce the output A AND $\overline{B}$; this may be expressed as A.$\overline{B}$, (following on from the alternative way of referring to AND as the 'logical product'). Then the two inputs A.$\overline{B}$ and C go into the OR-gate to produce A.$\overline{B}$ OR C or, as stated, (A.$\overline{B}$) + C, using the brackets to hold together the two parts of the output from the AND-gate. D is the name given to the final output from the circuit. Note how we may draw up the circuit diagrammatically *or* in algebraic form. In order to find the value of D a truth table needs to be produced which lists all possible values of A, B and C, as shown in Figure 15.10.

A	B	C	$\overline{B}$	A.$\overline{B}$	D
0	0	0	1	0	0
0	0	1	1	0	1
0	1	0	0	0	0
0	1	1	0	0	1
1	0	0	1	1	1
1	0	1	1	1	1
1	1	0	0	0	0
1	1	1	0	0	1

Figure 15.10   Truth Table

In the table, the additional columns represent the intermediate stages between input and output; $\overline{B}$ is tabulated since it is an input into the AND-gate and is derived directly from B; then since the AND-gate produces $A.\overline{B}$ which has to be input into the OR-gate later, we calculate $A.\overline{B}$ using the A and $\overline{B}$ columns and the rules for the AND operation. Finally having got $A.\overline{B}$ and C we can calculate D by using the two values as input to the OR-gate. It can at once be determined what the value of D will be for any given combination of input values A, B and C.

The objective in most practical situations is to design the logic circuit but, in order to do this with the maximum of efficiency it is necessary to use the Boolean algebraic statement of such a circuit, together with techniques to assist in its design.

Subsequently the algebraic expressions may require simplification and this in turn will lead to simpler circuits. To this end will be used De Morgan's Laws (section 15.4.2), Karnaugh Maps (section 15.4.3), Venn diagrams (section 14.8), etc, each of which can assist in such simplification and in clarifying the logic in the process.

Computers use logic circuits for a variety of reasons, including the control of input/output devices, access to storage, coders and decoders, arithmetic and other calculation processes, and in registers for purposes of control. In hardware circuits, logic gates are replaced by transistors serving exactly the same function; a number of such transistors could be replaced by a solid-state device, the integrated circuit (or IC) and these serve the same function as a number of gates so may be designed for specific applications. Since the integrated circuit can combine a million or so of these gates and since they can combine both storage and processing components, they are the 'chips' or low-cost microprocessors appearing in so many devices today; their advantages include reliability, small size and low power consumption.

15.3   A SIMPLE EXAMPLE – THE HALF ADDER

In the circuit illustrated in Figure 15.11, there are two inputs A and B and two outputs X and Y.

Note that A is 'split' so that its signal goes directly into *both* the

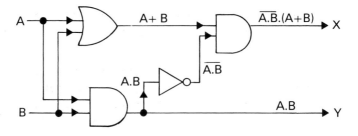

Figure 15.11   A Half Adder Circuit

first AND-gate and the OR-gate; this does *not* imply any reduction in its magnitude. The same is true also for B which is 'split' in a similar fashion. The circuit is annotated to show the on-going output at each stage as the gates are encountered; thus the output from the OR-gate is shown as A + B and the output from the NOT-gate appears as $\overline{A.B}$.

The first of the truth tables which appear in Figure 15.12 shows each of these intermediate outputs and defines their values, so

A	B	A+B	Y A.B	$\overline{A.B}$	X $\overline{A.B}.(A+B)$
0	0	0	0	1	0
0	1	1	0	1	1
1	0	1	0	1	1
1	1	1	1	0	0

Simplified Truth Table:

A	B	X	Y
0	0	0	0
0	1	1	0
1	0	1	0
1	1	0	1

Figure 15.12   Simplified Truth Tables

leading to the X and Y values. The second truth table is merely a simplification of the first, deleting reference to all but the two inputs and the two outputs.

Examination of the second truth table provides some interesting results, especially if the X and Y columns were to be reversed, since something remarkably similar was seen earlier in the book with binary arithmetic. This in fact gives us the basis for adding together two binary digits:

$$0 + 0 = 0$$
$$0 + 1 = 1$$
$$1 + 0 = 1$$
$$1 + 1 = 10 \text{ ie } 0 \text{ with a 'carry' of 1.}$$

It follows that A and B are the two binary digits to be added, that X is their 'sum' and that Y is the 'carry' to the next column. This circuit does not as yet go far enough, since it does not cope with such things as a 'carry' digit from a previous column; nonetheless, it is an important circuit. It is known as a *Half Adder* and will be encountered again in the next chapter as part of the *Full Adder*.

15.4   SIMPLIFICATION OF BOOLEAN EXPRESSIONS

15.4.1   Algebraic Laws

There are a large number of different simplifications that can be made to algebraic representations of logic circuits. The first batch of results which can be used in the pursuit of simplification stem from consideration of a single set A and its place within the Universal Set (Figure 15.13).

To appreciate these rules, consider that 1 is equivalent to the Universal Set and that 0 is equivalent to the empty set. Rule (i) states simply that the intersection of the empty set with A is just the empty set itself. In rule (ii) the union of A with the empty set simply gives A. Rule (iii) considers the intersection of A with the Universal Set; evidently the common ground between the two can only be A itself. In (vii) the intersection between A and its complement A' must be the empty set, whilst the union of these two (rule (viii)) is clearly seen as the Universal Set.

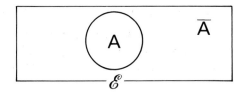

RULES : (i) $A \cdot O = O$

(ii) $A + O = A$

(iii) $A \cdot 1 = A$

(iv) $A + 1 = 1$

(v) $A \cdot A = A$

(vi) $A + A = A$

(vii) $A \cdot \overline{A} = O$

(viii) $A + \overline{A} = 1$

Figure 15.13   Algebraic Laws – Universal Set

The next diagram in Figure 15.14 involves two intersecting sets A and B and the four laws which emerge from considering their interrelationships. For example rule (ix) states that the intersection of A with the union of A with B, (that is the area common to both A and the union of A with B) must be just A; hence $A.(A+B) = A$. Similarly in (x) A.B represents the intersection of

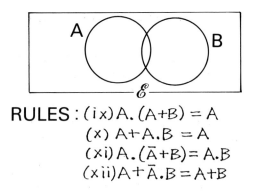

RULES : (ix) $A \cdot (A+B) = A$

(x) $A + A \cdot B = A$

(xi) $A \cdot (\overline{A}+B) = A \cdot B$

(xii) $A + \overline{A} \cdot B = A+B$

Figure 15.14   Algebraic Laws – Intersecting Sets

A with B; the union of this with A will just get A and so on.

The next set of rules, the *Commutative Laws* shown in Figure 15.15, imply simply that, when taking the union or the intersection of two sets it does not matter in which order they are considered.

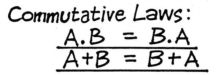

Commutative Laws:
$$A.B = B.A$$
$$A+B = B+A$$

Figure 15.15 Commutative Laws

The *Associative Laws* state that, when three propositions are involved, it does not matter which two are combined first; so that A.(B.C) or (A.B).C both provide equivalent ways of evaluating A.B.C (Figure 15.16).

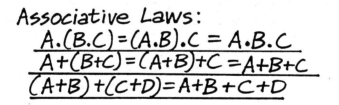

Associative Laws:
$$A.(B.C)=(A.B).C = A.B.C$$
$$A+(B+C)=(A+B)+C =A+B+C$$
$$(A+B)+(C+D)=A+B+C+D$$

Figure 15.16 Associative Laws

The *Distributive Laws* indicate that A.(B+C) can be evaluated by 'multiplying out' the bracketed expression to produce A.B + A.C; this parallels closely the conventional rules of algebra as introduced in Chapter 10. The relationship based upon A+B+C = 1 takes rather longer to deduce but relies upon earlier results and the reverse of the distributive law so that A.C + A.B = A.(C+B); then, since A+B+C = 1, it follows that B+C = $\overline{A}$, as well as using the A.$\overline{A}$ = 0 result. Note that B.C may be written as BC where there is no possibility of any confusion existing (Figure 15.17).

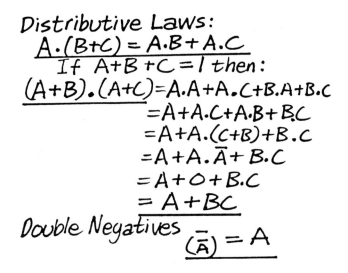

Figure 15.17   Distributive Laws

The 'double negative' statement implies quite simply that the complement of the complement of A is just A itself; it may seem obvious but it still repays stating.

15.4.2   De Morgan's Laws

Strictly speaking De Morgan's Laws are as much algebraic as those dealt with in the previous section but they have been kept separate since they have to be applied to a Boolean expression in three clearly defined stages, which are:

(a)   change every AND to OR and every OR to AND;

(b)   negate *all* variables;

(c)   negate the whole expression.

In the following examples (Figure 15.18) these rules are applied to two different problems, proving in the first case that $\overline{A+B} = \overline{A}.\overline{B}$ and in the second case the equivalent result that $\overline{A.B} = \overline{A}+\overline{B}$.

The next example (Figure 15.19) is slightly more involved and

eg (a) $\overline{A+B}$ is to be simplified
after (i) it becomes $\overline{A \cdot B}$
after (ii) it becomes $\overline{\overline{A} \cdot \overline{B}}$
after (iii) it becomes $\overline{A} \cdot \overline{B}$ $\therefore \overline{A+B} = \overline{A} \cdot \overline{B}$

eg (b) $\overline{A \cdot B}$ is to be simplified
after (i) it becomes $\overline{A+B}$
after (ii) it becomes $\overline{\overline{A} + \overline{B}}$
after (iii) it becomes $\overline{A} + \overline{B}$ $\therefore \overline{A \cdot B} = \overline{A} + \overline{B}$

Figure 15.18   Application of De Morgan's Laws to Simple Expressions

eg (c) Simplify $(\overline{A+B}) \cdot \overline{C}$
Firstly use De Morgan's laws to convert
$\overline{A+B}$ to $\overline{A} \cdot \overline{B}$ so we now have
$\overline{A} \cdot \overline{B} \cdot \overline{C}$. Next use De Morgan's laws
on this result to give $\overline{A+B+C}$
$\therefore (\overline{A+B}) \cdot \overline{C} = \overline{A+B+C}$
Likewise $\overline{A+B+C+D} = \overline{A} + \overline{B} + \overline{C} + \overline{D}$ etc.

Figure 15.19   Using De Morgan's Laws on a more Involved Expression

we deal with $(\overline{A+B})$ first using De Morgan's Laws but initially ignoring the presence of $\overline{C}$. Only when $(\overline{A+B})$ has been changed to $\overline{A} \cdot \overline{B}$ is $\overline{C}$ reintroduced.

15.4.3   Karnaugh Maps

There are physical similarities between Karnaugh Maps and Venn diagrams, since they can both be used to represent Boolean expressions algebraically. It is possible to use Karnaugh Maps to simplify expressions with as many as six variables although in this chapter they will be restricted to two- and three-variable cases, so dealing only with circuits in which there are only two or three inputs. The process can however, be used to deal with far more complex simplifications.

The diagram in Figure 15.20 shows the two-input Karnaugh Map; with A and $\overline{A}$ along the top and B and $\overline{B}$ down the side, every one of the four cells is immediately defined by relation to the elements at the top and at the side hence as AB, $\overline{A}$B, A$\overline{B}$ and $\overline{A}\overline{B}$ respectively.

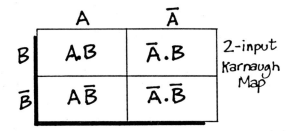

Figure 15.20   2-input Karnaugh Map

In the next diagram in Figure 15.21 the whole of A (ie AB + A$\overline{B}$) is shown shaded in the left-hand illustration and the whole of B (ie AB + $\overline{A}$B) can be seen shaded in the diagram on the right. It is possible to identify any one of the elements A, $\overline{A}$, B or $\overline{B}$ by reference to appropriate shading of components within this map.

If given a Boolean expression such as AB + $\overline{A}$B + $\overline{A}\overline{B}$, as shown

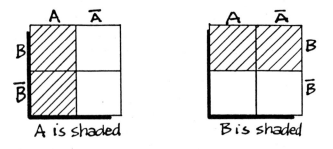

Figure 15.21   Shaded Karnaugh Maps

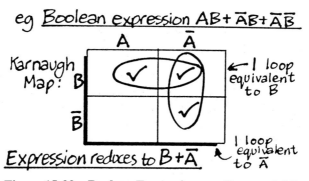

Figure 15.22   Boolean Expression as a Karnaugh Map

in Figure 15.22, each of the three constituent parts can be ticked on the Karnaugh Map. Next, loops are drawn around each pair of adjacent squares that are ticked, or any isolated square that is not so adjacent. Each loop now identifies the variable present in the simplification of the Boolean expression. In the diagram the loops identify B and $\overline{A}$ and the fact that the loops overlap is totally irrelevant; hence we find that AB + $\overline{A}$B + $\overline{A}\overline{B}$ is capable of simplification to B + $\overline{A}$ and this simpler form has exactly the same effect. Thus any circuit represented in this way could be simplified by another which does exactly the same but which requires far less circuitry.

It is now possible to proceed to the three-variable form of the Karnaugh Map. Since there can only be two sides to the rectangle drawn here we must let A and A go down one side whilst the top edge is made up of the four components BC, BC, BC and BC (which make up all the possible combinations of B and B with C and C). Again, each of the eight squares identifies one of the eight possible combinations involving A or A with B or B with C or C (Figure 15.23). A itself can be identified as being the whole of the top four squares whilst $\overline{A}$ is given by the bottom four. $\overline{B}$ is identifiable as being the central four squares since

$$(\overline{B}C + \overline{B}\overline{C})(A + \overline{A}) = \overline{B}(C+\overline{C}) = \overline{B}$$

and B is given by the two vertical squares on the left hand end combined with the two vertical squares at the right hand end. Likewise C could be identified as the left hand end block of four squares and C as the right hand end block of four. Other items can

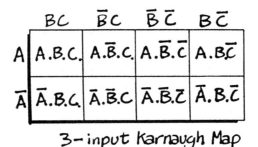

3-input Karnaugh Map

Figure 15.23   3-input Karnaugh Map

be defined by taking two squares at a time.

To simplify the Boolean expression
$$ABC + AB\bar{C} + A\bar{B}C + \bar{A}BC + \bar{A}\bar{B}C$$
the five units on the Karnaugh Map are ticked and loops of four
squares or two squares are drawn as appropriate (Figure 15.24).

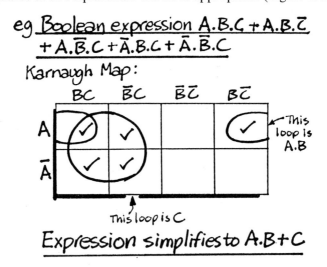

**Figure 15.24   Boolean Expression Simplified Using a Karnaugh
Map**

Since the left-hand end block of four squares is totally contained
the presence of C in the simplification is easy to establish, but note

also that ABC + AB$\overline{\text{C}}$ 'wraps round the back' to simplify to AB. Hence the whole expression reduces to AB + C, a considerable reduction from the original.

Karnaugh Maps are most valuable in dealing with expressions in which there are a number of items linked together by ORs.

15.5   EXERCISES

1.   Copy and complete the following truth table:

A	B	A.B	A+B	(A+B) . (A.B)
0	0	0	0	0
0	1	0	1	0
1	0	0	1	0
1	1	1	1	1

2.   With reference to the following circuit:

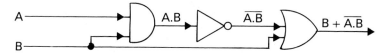

copy and complete the truth table:

A	B	A.B	$\overline{\text{A.B}}$	B + $\overline{\text{A.B}}$
0	0	0	1	1
0	1	0	1	0
1	0	0	1	1
1	1	1	0	1

3. With reference to the following circuit:

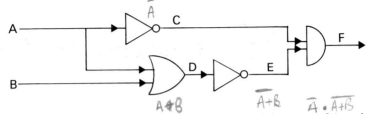

write down the outputs at C, D, E and F, each in terms of A and B, and produce a truth table to define F for every combination of values of A and B.

4. With reference to the following circuit:

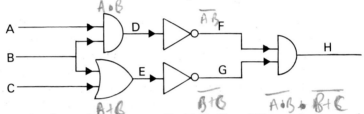

write down the outputs at D, E, F, G and H each in terms of A, B and C. Produce a truth table to define F for every combination of values of A, B and C.

5. Use De Morgan's Laws to simplify each of the following:

(i) $\overline{A} + \overline{B} + \overline{C}$

(ii) $(\overline{A} + \overline{B}) . (\overline{B} + \overline{C})$

(iii) $\overline{A}.\overline{B} + \overline{B}.\overline{C}$

(iv) $(\overline{A.B}) + (\overline{B.C}) + (\overline{A.C})$

6. Use a Karnaugh Map to simplify each of the following:

(i) $A.B + A.\overline{B} + \overline{A}.\overline{B}$

(ii) $A.B + \overline{A}.B$

(iii) $A.B.C + \overline{A}.B.\overline{C} + A.B.\overline{C} + \overline{A}.B.C$

(iv) $\overline{A}.\overline{B}.C + \overline{A}.\overline{B}.\overline{C} + A.B.\overline{C} + \overline{A}.B.\overline{C}$

(v) $A.\overline{B}.\overline{C} + A.B.\overline{C} + A.B.C + A.\overline{B}.C + \overline{A}.\overline{B}.C$

16 Applications of Logic

16.1 THE FULL ADDER

When the half adder was introduced in the previous chapter, it was noted that it had limited use if it was used on its own since it was incapable of handling a 'carry' from a previous addition. Thus methods need to be explored of overcoming this drawback. The solution is the full adder which, by combining two half adders and an OR-gate, can cope with any two inputs as well as the previous 'carry' and produce a 'sum' and a 'carry' forward. This is clearly shown in Figure 16.1 below.

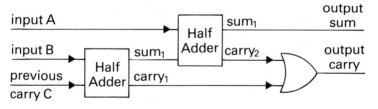

Figure 16.1   A Full Adder

Note how the half adders appear now only as blocks; as circuits get more involved so certain well-defined components, such as the half adder, are replaced in this way to reduce the overall complexity of the diagram. Were this not done the full adder would have a diagram consisting of four AND-gates, three OR-gates and two NOT-gates and so would be unnecessarily complicated.

The truth table in Figure 16.2 defines the 'sum' and 'carry'

FULL ADDER

INPUTS			OUTPUTS	
A	B	C	SUM	CARRY
O	O	O	O	O
O	O	I	I	O
O	I	O	I	O
O	I	I	O	I
I	O	O	I	O
I	O	I	O	I
I	I	O	O	I
I	I	I	I	I

Figure 16.2   Full Adder Truth Table

outputs from the full adder for the three inputs A, B and the previous carry C.

In the same way it is possible to illustrate how two full adders can be used to add together two 2-bit numbers. Note again how the full adder, in its turn, is replaced by a block component in this diagram. The 'first' full adder at the right-hand side is adding the 'units' digits together and requires a zero carry from the previous (non-existent) operation. However the 'second' full adder at the left, which adds the 2^1 digits together takes in the genuine carry from the work of the first full adder. The carry from this second full adder acts as the 2^2 digit in the result 101. If we had to carry out the addition of two 12-bit numbers then 12 full adders would have to be harnessed together in a manner quite similar to that used in Figure 16.3

16.2   FLIP-FLOPS

This logic circuit is used for the purpose of data storage. There are two inputs R (for reset) and S (for set) and two outputs Q and $\overline{Q}$.

PARALLEL ADDITION OF
TWO 2-BIT NUMBERS

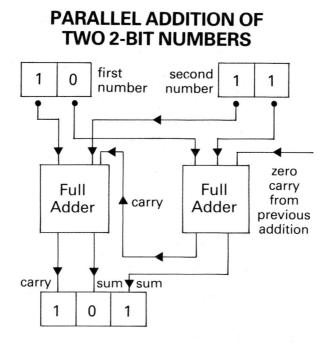

Figure 16.3   Parallel Addition of Two 2-bit Numbers

For any given initial value of Q the values of R and/or S may cause Q to change. However, this change (if any), is also dependent upon the initial value of Q. It follows therefore that the output Q is fed back into the input side.

If Q is initially 0 (ie off) then *resetting* it (R = 1) can have no effect, but *setting* it (S = 1) must turn it on, giving it a value of 1. If Q is initially 1 (ie on) then setting it (S = 1) can have no effect since it is already set but resetting it (R = 1) will change it back to 0. It is not possible to both set *and* reset the input Q at the same time so we get the two 'not feasible' entries in the truth table in Figure 16.4.

Note carefully in the circuit diagram that both Q and $\overline{Q}$ are used to re-input into the OR-gates. This reset-set flip-flop is also referred to as a *bistable*.

R.S. FLIP-FLOP OR BISTABLE

R	S	Initially Q	Now Q
0	0	0	0
1	0	0	0
0	1	0	1
1	1	0	Not Feasible
0	0	1	1
1	0	1	0
0	1	1	1
1	1	1	Not Feasible

NB: Both Q and $\overline{Q}$ are required for outputs

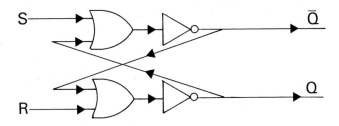

Figure 16.4   Reset-set Flip-flop or Bistable

16.3   LIGHT SWITCHES

Next is a fairly common type of industrial application for a logic circuit (Figure 16.5). In an office there are three separate light switches that control the same light but from different places; each switch may be off (0) or on (1). The light itself must be on (1) if there are an odd number of switches on but off (0) otherwise. Clearly, if the light is already on nobody would touch a switch unless it was to turn the light off. The truth table describes the possible values of the three switches A, B and C and the required conditions of the output (the light) in each case.

A	B	C	output
0	0	0	0
0	0	1	1
0	1	0	1
0	1	1	0
1	0	0	1
1	0	1	0
1	1	0	0
1	1	1	1

Figure 16.5 Light Switch Truth Table

Examination of the table indicates that *there is* an output (ie the light is on) if $A = 0$, $B = 0$ and $C = 1$ or if $A = 0$, $B = 1$, $C = 0$, etc. These four cases are described as $\overline{A}\overline{B}C$, $\overline{A}B\overline{C}$, $A\overline{B}\overline{C}$ and ABC respectively (since $\overline{A}\overline{B}C$ implies that $A = 0$, $B = 0$, $C = 1$) so that the output can be described by the Boolean expression which is $\overline{A}\overline{B}C + \overline{A}B\overline{C} + A\overline{B}\overline{C} + ABC$. This is next illustrated on the Karnaugh Map in Figure 16.6 which indicates that there is no possibility of simplification:

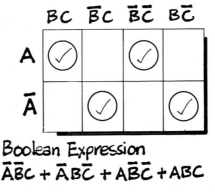

Boolean Expression
$$\overline{A}\overline{B}\overline{C} + \overline{A}B\overline{C} + A\overline{B}\overline{C} + ABC$$

Figure 16.6 Boolean Expression of Light Switch Truth Table Simplified into a Karnaugh Map

Now that the nature of the circuit is known it is possible to develop the drawing, in Figure 16.7 which shows the four components created separately and fed into a four-input OR-gate, as well as the three inputs A, B and C.

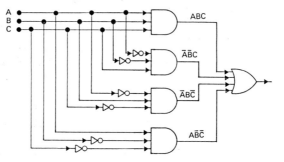

Figure 16.7   Four-input Or-gate

The example in Figure 16.7 illustrates the technique for deriving a circuit from a knowledge of the output(s) that are required for various states of the inputs. In a similar manner other circuits may be derived even if they may prove to be more involved than this.

16.4   NAND-GATES AND NOR-GATES

Quite frequently additional gates may arise in the construction of a logic circuit, of which two of the most common are the NAND-gate and the NOR-gate. These have their own symbols and their own truth tables. The NAND-gate is really a NOT/AND gate so in effect it consists of an AND-gate with a NOT-gate immediately afterwards to invert the result of the AND-gate. Both the symbol in use and the truth table in Figure 16.8 reflect this relationship.

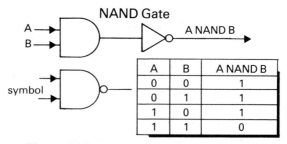

A	B	A NAND B
0	0	1
0	1	1
1	0	1
1	1	0

Figure 16.8   Nand-gate with Truth Table

In a similar manner the NOR-gate is in effect an OR-gate with a NOT-gate immediately afterwards to invert its effect. The symbol and the truth table for the NOR-gate are shown in Figure 16.9.

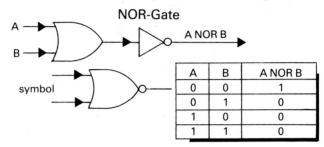

Figure 16.9   Nor-gate with Truth Table

Anything which can be achieved by use of AND, OR and NOT gates can equally well be redesigned so as to use NAND and NOR gates only. The diagram in Figure 16.10 illustrates a half adder which uses NAND gates only.

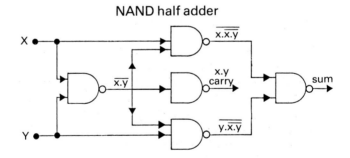

Figure 16.10   Nand Half Adder

16.5   PROGRAMMING RELEVANCE

Although all the examples so far given on the application of logic have related to the hardware aspects of computing, this is by no means the only area of use for such methods, even if the derivation of circuits as such may have little relevance in the context of software. Examples of using logic in a programming context can easily be found both as applied to high level languages and in the context of machine code.

Consider a language such as Cobol and the use of a compound condition statement such as:

IF TAX-VAL > 25 AND CODE-X NUMERIC OR
GROSS-TOT NEGATIVE PERFORM X-PARA.

Here there are three elementary conditions TAX-VAL > 25 and CODE-X NUMERIC and GROSS-TOT NEGATIVE; calling these conditions c-1, c-2 and c-3 respectively the condition can be restated as:

IF c-1 AND c-2 OR c-3 PERFORM X-PARA.

The problem is to determine whether the compound condition c-1 AND c-2 OR c-3 is true or false (so that X-PARA is or is not executed) and this in turn depends upon the truth or falsehood of each of the three elementary conditions. In this respect there are therefore the three inputs c-1, c-2 and c-3 and the logic circuit based upon the appropriate combination of these inputs. Many other such situations exist in programming, requiring a knowledge of how to process such compound statements. The rules, insofar as they are specific to programming (which they really are not, being far more general), would be dealt with in any course in programming and in particular on the use of compound conditions. Even the use of NOT to invert a condition is normal practice in programming situations.

Likewise when working at machine-code level situations may be encountered in which the bit-patterns held in two storage locations have to be compared for the purpose of carrying out logical operations upon them. We may wish to mask out a portion of a word so that attention can be focused on, for example, a specific field within a program instruction. Alternatively we may wish to set or to reset certain bits in a word, acting as switches, a task achieved by the use of logic operations. Sometimes we may need to ensure that the value of a particular word oscillates between two alternate values, again by the imposition of some logical process upon the bits.

Logic is therefore not solely the province of the hardware

designer; it has a large number of quite distinct programming applications. Although they are not dwelt on here in depth this is only because the subject more properly belongs to courses in the respective programming languages, be they at the level of machine-code or at the high-level stage.

16.6   EXERCISES

1. The company safe has three locks. The keys are held, one each, by the manager and his two assistant managers. Devise a logic circuit that will allow the safe to be open (=1) if the manager together with one or both of his assistant managers use their keys but to remain locked otherwise. *Hint:* draw up a truth table listing all possible combinations of the presence (=1) or the absence (=0) of each of the three keys, then simplify the result by use of a Karnaugh Map or otherwise before drawing up the circuit.

2. Design a circuit to control the starting (=1) of a data recording device, which requires *all three* of the following to make it work:

    (i)   data recording device is switched on (=1);

    (ii)   it is connected to the modem (=1);

    (iii) the program asks for data recording (=1).

3. Three people at a meeting can each vote by pressing a button (=1) or not pressing it (=0), this latter implying a vote against the motion. A decision is taken in favour of a motion (=1) if *any two or more* buttons are pressed. Design a circuit to control this.

4. Four inputs represent the binary quantities 1, 2^1, 2^2, 2^3. Design a circuit which will allow output (=1) if the inputs identify a decimal digit in the range 0 to 9 only.

17 Arrays

17.1 INTRODUCTION

Arrays occur quite naturally in programming and in the use of a number of standard packages, notably including spreadsheets. They may be used in commercial work for holding a reference table, containing for example, a variety of different commission rates or delivery charges; alternatively they may hold a small file which has to be processed internally in some way, possibly for the purpose of sorting it into some defined sequence. In scientific applications too they are common, although tending to contain numeric data, in which case the alternative name of matrices (singular 'matrix') may be used. Arrays also occur in the way in which the machine processes instructions or handles remote enquiries, when data may be stored in stacks or queues, etc which may very conveniently be represented by arrays.

17.2 DEFINITION

If a list of values were taken, most typically of the same variable such as the following list of customer account numbers:

P328   A714   F305   A286   P394   R113   F219   A664   F317

then it would be possible to define a specific value by reference to its position in the list so we could refer to item 4 in this list rather than to A286. If the whole of this list (of nine items) was given the name C, then we would be able to refer to each item by the use of the name C and the position in the list; hence C_6 could be written instead of R113, etc. This is an example of a list or one-dimensional array (sometimes called a linear array) and we refer to C_6 as a *subscripted variable,* 6 being the subscript; in this particular example it only makes sense of course, if the subscript is a number in the range 1 to 9 inclusive. The value of doing this is manifold; firstly it reduces the number of different variables needed within any one problem, since C in the example caters for all the customer account numbers with no need to have 9 different names, a point which may not be too important in the abstract, but which is of value in the case of certain programming languages which possess a very restrictive set of allowable variable names. The second advantage is far more significant; it allows a set of interrelated data values to be grouped together, and subsequently to be manipulated. The grouping may be, as in this case, the customer account numbers for many customers, or it might be a set of fields relating to the same record.

This concept can be extended through into the two-dimensional array as illustrated in Figure 17.1.

6	5	13	0	8	-3	6
2·9	4	8	-1	(2)	8·1	-4
18	15	-3·7	2	2	3·8	6
-8	0	(14)	5	3·8	2·9	-2
3	8·1	6	3	17	(8)	4·7

T(2,5)
T(5,6)
T(4,3)
NB Using program notation

Figure 17.1 Two-dimensional Array or Table

In this case the array is frequently called a table (or a matrix if it only contains numeric data items) and it can be thought of as being subdivided into rows and columns. Hence, to identify a specific value in the table it will be necessary to refer both to its row and its column as well as to the name given to the table. The table in Figure 17.1 is called T; it consists of 5 rows and 7 columns, so enabling it to hold 35 values. Some values have been referred to specifically such as T(5,6) which is the item in the row number 5, column number 6 of table T and this has the value 8 ie $T(5,6) = 8$. The notation used here is that commonly found in a program and whereas we might use $T_{5,6}$ in a rather more abstract sense this does not suit programming languages, most of which cannot cope with 'below the line' subscripts (*both* the 5 and the 6 are subscripts in this case).

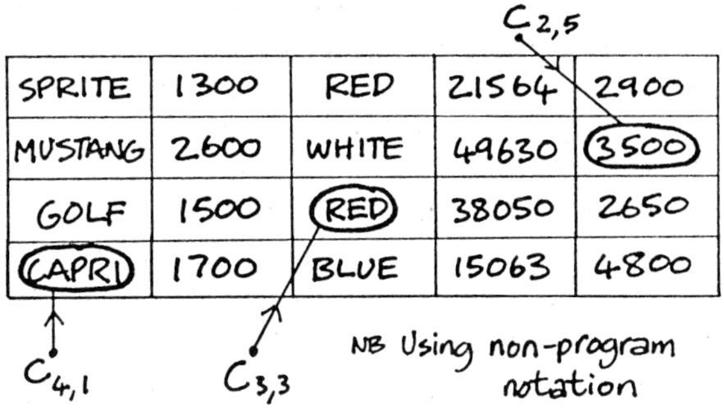

Figure 17.2   Two-dimensional Array or Table C

In the second case (Figure 17.2) table C contains a mixture of numeric and alphabetic data, but the notation used remains exactly the same, noting again that $C(2,5)$ might be used in a program instead of the $C_{2,5}$ used here to refer to the value of 3500.

Although not likely to be very concerned with arrays other than in the cases of lists and tables, it is possible to extend the concept into three or more dimensions. The diagram in Figure 17.3 illustrates a three-dimensional case, the third subscript identifying the 'depth' into the structure, thus $T(1,5,5)$ or $T_{1,5,5}$ identifies the element in row 1, column 5 and 5 'deep'.

The three-dimensional case is not too difficult to visualise, although a little tedious to draw. Many languages do in fact allow more than three dimensions, but these are impossible to illustrate so readily in a drawing and, for many people, quite difficult to appreciate.

It should be stressed that, although certain arrays have been referred to as being two- or three-dimensional, this is the way in which they are viewed, rather than being necessarily the way in

THREE - DIMENSIONAL ARRAYS

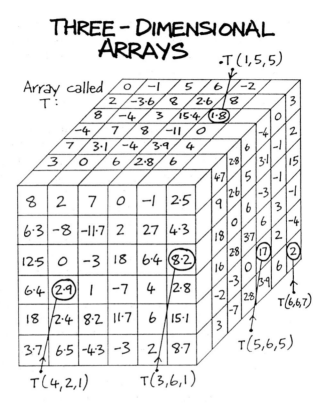

Figure 17.3   Three-dimensional Array T

which the machine stores them. Thus it is quite likely that a machine may store a 5 row by 7 column array as a linear array of 35 items so that the second row may in fact be items 8 to 14 in that linear array; this makes no difference from the point of view of the programmer who can continue to *regard* the array as being two-dimensional and who then allows the software to convert between the stored data and the way in which he or she 'sees' it as being stored.

Apart from the use of numeric subscripts variables may also be used as subscripts; thus we may refer to L_X or $L(x)$ in the case of a list, or to $T(X, Y)$ or $T_{X,Y}$ in the case of a table, hence enabling values to be assigned to X (and Y) at the time that processing takes place. In the case of a 9-element list C whose elements are C_1, C_2, etc the general element may therefore be referred to as C_X and X may take the values from 1 to 9, possibly within a program loop, to identify each element in turn.

17.3   CONSTRUCTING ARRAYS FROM RAW DATA

An array is, of course, not just a way of holding a set of disorganised data together. It possesses some sort of definable structure and in the one-dimensional case it is most likely that all the elements are of the same type; for example, they might each be an examination mark in a given subject for several different candidates, the age of a person in a particular group of people or the length of time between consecutive interrupts. The order in which the items occur in the list may be the natural order in which they have been selected or in which they occur as functions of time or just by chance. Subsequently we may wish to restructure the array by a process such as that of sorting the elements into ascending numeric or alphabetic order. The diagram in Figure 17.4 shows an example of the way in which data may be read by a Basic program for the purpose of storing it into an array A. Note that this shows also the array before and after the execution of the instructions. However, it is certainly possible that the array A may have not been empty initially but that it may have held other data. Note also the point made, which is really of more importance to the programming aspects, that the array may well be assumed to possess a zero-th element such as A(0) as well as the others.

In the two-dimensional case it may be necessary to pay more regard to the way data is loaded in since it is more than likely that this time a mixture of fields will be involved. Typically, it could be that each row identifies an individual record (a company employee for example), and that each element in that row is a field in such a record, the name, date-of-birth, pay-to-date, tax-to-date, date of joining the company and so on. Hence within any one *column* there will be a set of names or of dates-of-birth or of pay-to-date for all the employees. It is totally realistic that the data will be a mixture of numeric and alphabetic data, although this can be problematic with some programming languages; Cobol readily accepts such a mixture whereas many dialects of Basic require that

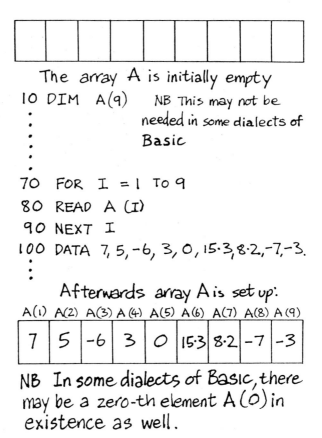

The array A is initially empty

```
10 DIM  A(9)    NB This may not be
 .               needed in some dialects of
 .               Basic
 .

70  FOR  I = 1  TO 9
80  READ  A (I)
90  NEXT  I
100 DATA  7, 5, -6, 3, 0, 15·3, 8·2, -7, -3.
 .
 .
```

Afterwards array A is set up:

A(1)	A(2)	A(3)	A(4)	A(5)	A(6)	A(7)	A(8)	A(9)
7	5	-6	3	0	15·3	8·2	-7	-3

NB In some dialects of Basic, there may be a zero-th element A (0) in existence as well.

Figure 17.4   Constructing an Array from Raw Data

the whole of an array is *either* numeric *or* is in alphanumeric format. In such a case it may be necessary to convert the numeric elements into strings by the use of the STR$ function so making the whole array alphanumeric and then using the VAL function to convert the strings back to numeric values as appropriate later.

In Figure 17.5 a two-dimensional array is shown holding the examination results of six students, together with two distinct ways of loading the data into the table. In the first, data is loaded, one student record at a time, ie on a 'row by row' basis whereas in the

STUDENT NUMBER	MATHEMATICS	HISTORY	LATIN	SCIENCE
1307	46	58	24	49
1314	71	42	54	83
1318	24	17	20	19
1321	83	46	53	78
1327	75	58	62	88
1334	40	32	6	47

Coding: eg 1

```
10 DIM R(6,5)
   :
30 FOR I = 1 TO 6
40 FOR J = 1 TO 5
50 READ R(I,J)
60 NEXT J
70 NEXT I
80 DATA 1307, 46, 58,24,49, 1314, 71, 42,etc.
```

Coding: eg 2

```
10 DIM R(6,5)
   :
30 FOR J = 1 TO 5
40 FOR I = 1 TO 6
50 READ R(I,J)
60 NEXT I
70 NEXT J
80 DATA 1307,1314,1318,1321,1327,1334,46,71,etc.
```

Figure 17.5 Two-dimensional Array and Examples of Loading Data

second case it is loaded on a 'column by column' basis, thus dealing with each subject in turn. The method to be employed depends upon circumstances although the former approach is frequently the more natural.

More involved forms of array such as the three-dimensional, can be created in a precisely comparable manner. The programming may prove to be slightly more involved although the principles remain exactly the same; developing the examination result approach, it is not too difficult to perceive a 6-row by 4-column structure (deleting the student number column) with values going 'into' the paper which define the marks achieved by a particular candidate in a specified subject for each question on the examination paper.

17.4 EXERCISES

1. By reference to the following list, L:

    17.1   12.3   6.8   9.7   5.4   7.8   5.3   11.1   9.6   10.4

    what are the values of (i) L_3;

    (ii) L_7;

    (iii) L_{11}?

2. With reference to the following table, T:

Fred	Nancy	Liang	Lakbir	Suzanne	Li	David
Andrew	Rukshini	Scott	Gina	Fatima	Jan	Ho
Harriet	Sukvinder	Ong	Suzie	Danielle	Sharjah	Maria
Uppal	Daipal	Tony	Petra	Franz	Ram	Drogo
Anna	Zsu-Zsu	Yola	Philip	Atwal	Hannah	Hong

    what are the values of (i) $T_{2,6}$;

    (ii) $T_{3,1}$;

    (iii) $T_{1,3}$;

    (iv) $T_{5,4}$?

    How may we refer to (v)   Daipal?

    (vi)   Franz?

17.5 ADDING UP ROWS OR COLUMNS

When dealing with the processing of an array it is quite common to find that there is a need to add up the elements in a given row or column. Consider the example in the previous section on student examination results; if all the elements in column 4 were added up and then the result divided by 6 the average mark for Latin would be obtained. Similarly if all the elements were added up in row 5, columns 2 to 5 only, and then divided by 4 we would find the average mark scored by candidate 1327. The coding in Basic for both cases is illustrated in Figure 17.6.

eg. To calculate the average Latin mark:
```
150  L = 0
160  FOR  I = 1 TO 6
170  L = L + R (I, 4)
180  NEXT  I
190  PRINT 'AVERAGE LATIN MARK IS'; L/6
```
With the data used, this will result in:

AVERAGE LATIN MARK IS 36·5

                     being output

eg To calculate the average examination
    :   mark for candidate 1327:
```
250  C = 0
260  FOR  J = 2 TO 5
270  C = C + R (5, J)
280  NEXT  J
290  PRINT 'CANDIDATE'; R(5,1); 'AVERAGED'; C/4
```
With the data used, this will result in:

CANDIDATE 1327 AVERAGED 70·75

                  being output

Figure 17.6   Processing an Array in Basic

A point to note however, is that addition of figures in this way is only sensible if it involves adding like with like so that the result of doing so is itself sensible. We may agree to total up all the candidate examination numbers to obtain a hash total for batch control purposes, but it would not be sensible to add together a candidate's examination number to his or her marks in four examinations, since no meaning whatsoever could be attached to the result.

17.6   ASSIGNMENT OF VALUES

Rather than 'reading in' a set of variable quantities into an array, specific pre-determined values may be assigned to their elements. The illustration in Figure 17.7 identifies three different cases; in the first the intention is to 'zeroise' each of the 15 elements in the linear array L, which could, as with the other examples, have been achieved by the use of READ . . . DATA . . . statements but the coding used here is far more efficient for the purpose. The second case is not dissimilar but the third reminds us that we may only wish to assign values to *certain* elements rather than to all:

eg Assign the value 0 to each item in the one-dimensional array L which has 15 elements $L(1)$ to $L(15)$:

```
10 DIM L(15)
   :
70 FOR I = 1 TO 15
80 L(I) = 0
90 NEXT I
```

eg Assign the value 1 to $M(1)$, 2 to $M(2)$, 3 to $M(3)$, etc up to 30 to $M(30)$ for the 30 element list M:

```
10 DIM M(30)
    :
130 FOR I = 1 TO 30
140 M(I) = I
150 NEXT I
    :
```

eg Assign the value $2^3 - 2^2 + 1$ to $K(2)$, $3^3 - 3^2 + 1$ to $K(3)$, etc up to $17^3 - 17^2 + 1$ to $K(17)$ for the 17 element list K ignoring the value of $K(1)$:

```
10 DIM K(17)
    :
210 FOR T = 2 TO 17
220 K(T) = T↑3 - T↑2 + 1
230 NEXT T
    :
```

Figure 17.7   Assignment of Values

17.7   ADDING AND SUBTRACTING ARRAYS

Just as it is possible to add or subtract single numbers so too we can add or subtract arrays which are structured collections of such numbers in almost identical manner. However, there are a number of provisos. Firstly the result of such addition or subtraction must be sensible; just as we would not expect to add together the tax-paid-to-date for an employee and the cost of an order from a customer so too care must be shown with regard to which arrays are being added or subtracted. Secondly the arrays involved must be of exactly the same format as one another; two arrays may be added together if they have precisely the same number of columns as one another as well as the same number of rows, but not otherwise (Figure 17.8). It (almost) goes without saying that it is possible only to add or subtract arrays if they hold numeric rather than alphabetic values.

Addition of 2 Arrays

$$
\begin{pmatrix} 7 & 5 & 9 \\ 4 & 0 & 6 \\ -3 & 2 & 5 \\ 1 & 4 & 2 \end{pmatrix} + \begin{pmatrix} 4 & 9 & 6 \\ 2 & 8 & -4 \\ 0 & 6 & 9 \\ -3 & 1 & 3 \end{pmatrix}
$$

$$
= \begin{pmatrix} 11 & 14 & 15 \\ 6 & 8 & 2 \\ -3 & 8 & 14 \\ -2 & 5 & 5 \end{pmatrix}
$$

Figure 17.8   Addition of Arrays

The Figure shows the case of the addition of two arrays, having 4 rows and 3 columns each. Note that, to achieve this, each element in the first array is added to its corresonding element in the second array to produce a similar element in the result, thus in row 1 column 1, 7 + 4 = 11.

The second example illustrates the subtraction of two arrays (Figure 17.9), each of which has 4 rows and 2 columns; once again

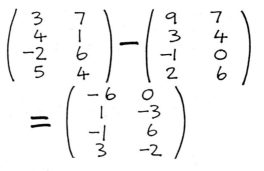

$$\begin{pmatrix} 3 & 7 \\ 4 & 1 \\ -2 & 6 \\ 5 & 4 \end{pmatrix} - \begin{pmatrix} 9 & 7 \\ 3 & 4 \\ -1 & 0 \\ 2 & 6 \end{pmatrix}$$

$$= \begin{pmatrix} -6 & 0 \\ 1 & -3 \\ -1 & 6 \\ 3 & -2 \end{pmatrix}$$

Figure 17.9   Subtraction of Arrays

the subtraction relates to corresponding elements, thus in row 3 column 2 is $6 - 0 = 6$.

17.8   SCALAR MULTIPLICATION

If we were to add an array to itself the result would be the creation of a new array in which every element was exactly twice what it had been before; alternatively what has been done is multiply the whole array by the (scalar) quantity 2. This process can be done for any scalar multiplier so that if we multiply an array by 5 the result is an array of precisely the same number of rows and columns as before but in which each element is five times larger. In the illustration given in Figure 17.10, an array M has been multiplied by 4.

If array M = $\begin{pmatrix} 6 & 3 & 8 & 7 & 5 \\ 2 & -1 & 0 & 4.1 & 3 \\ 9 & 5 & -7 & 2.1 & 6 \\ 4 & -3 & 0 & 2 & 8.5 \end{pmatrix}$

then 4M will be equal to:

$\begin{pmatrix} 24 & 12 & 32 & 28 & 20 \\ 8 & -4 & 0 & 16.4 & 12 \\ 36 & 20 & -28 & 8.4 & 24 \\ 16 & -12 & 0 & 8 & 34 \end{pmatrix}$

Figure 17.10   Scalar Multiplication of an Array

The second illustration (Figure 17.11) involves a combination of scalar multiplication with addition. Note that the final result is of exactly the same format as each of the two arrays started with.

$$6\begin{pmatrix} 3 & 2 & 1 \\ 5 & 1 & -3 \end{pmatrix} + 4\begin{pmatrix} 3 & 1 & -2 \\ -4 & 0 & 6 \end{pmatrix}$$

$$= \begin{pmatrix} 18 & 12 & 6 \\ 30 & 6 & -18 \end{pmatrix} + \begin{pmatrix} 12 & 4 & -8 \\ -16 & 0 & 24 \end{pmatrix}$$

$$= \begin{pmatrix} 30 & 16 & -2 \\ 14 & 6 & 6 \end{pmatrix}$$

Figure 17.11   Scalar Multiplication with Addition

17.9   MULTIPLICATION OF ARRAYS

It is possible to multiply two arrays together just as one array can be multiplied by a scalar. However, there are rules which restrict what can be done; in order to multiply out AB when A and B are two arrays, then:

(i)   the number of columns in A must be exactly equal to the number of rows to be found in B;

(ii) if rule (i) is not obeyed we *cannot define* the product AB.

An important outcome of this is that the order of multiplication is now of vital importance, which would not have been the case if two numbers such as 7 and 58 had been multiplied, as 7×58 is exactly the same as 58×7. If A is a 5 row by 4 column array and B is 4 by 3, then AB is defined (and will have 5 rows and 3 columns) but BA is not defined at all since the number of columns in B (3) is not the same as the number of rows in A (5). If A is a 2×3 array and B is a 3×2 array then AB is defined and will produce a 2×2 result; BA is also defined but will produce a 3×3 result.

Hence it can be generalised to suggest that AB = BA is *not generally true* in the case of array multiplication.

The method of achieving the product AB is seen by reference to the illustration which appears in Figure 17.12.

$$A = \begin{pmatrix} 5 & 1 & 4 \\ -3 & 6 & 7 \end{pmatrix} \quad B = \begin{pmatrix} 4 & 6 \\ 3 & 9 \\ -7 & 5 \end{pmatrix}$$

$$\therefore \underline{AB} = \begin{pmatrix} 5.4 + 1.3 + 4.-7 & 5.6 + 1.9 + 4.5 \\ -3.4 + 6.3 + 7.-7 & -3.6 + 6.9 + 7.5 \end{pmatrix}$$

$$= \begin{pmatrix} 20 + 3 - 28 & 30 + 9 + 20 \\ -12 + 18 - 49 & -18 + 54 + 35 \end{pmatrix}$$

$$= \begin{pmatrix} -5 & 59 \\ -43 & 71 \end{pmatrix}$$

Figure 17.12 Multiplication of Arrays

To find the element to go into row 1 column 1 of the result it is necessary to multiply together the row elements from row 1 of A by the column elements out of column 1 of B, leading to 5 × 4 + 1 × 3 + 4 × −7 and then total up the products 20 + 3 − 28 to give −5 as the result. In the same way, to find the element to go into row 1 column 2 of the result multiply together the row elements from row 1 of A and the column elements from column 2 of B and so on. The illustration in Figure 17.13 shows the calculation of the product BA.

Array multiplication is only valid if the result has sensible meaning. In the example given in Figure 17.14 the row-matrix A consists of the number of each of three items ordered, thus 7 of item-1, 6 of item-2, etc whilst the column-matrix B gives the prices of each of the three items, so £16 for each item-1, £8 for each item-2 and so on. Hence AB gives the total cost of the order (7 articles at £16 each plus . . .) whereas BA, although calculable, is meaningless.

$$\underline{\underline{BA}} = \begin{pmatrix} 4.5+6.\overline{3} & 4.1+6.6 & 4.4+6.7 \\ 3.5+9.\overline{3} & 3.1+9.6 & 3.4+9.7 \\ -7.5+5.\overline{3} & -7.1+5.6 & -7.4+5.7 \end{pmatrix}$$

$$= \begin{pmatrix} 2\phi-18 & 4+36 & 16+42 \\ 15-27 & 3+54 & 12+63 \\ -35-15 & -7+30 & -28+35 \end{pmatrix}$$

$$= \begin{pmatrix} 2 & 40 & 58 \\ -12 & 57 & 75 \\ -50 & 23 & 7 \end{pmatrix}$$

Figure 17.13   Calculation of the Product BA

$$A = \begin{pmatrix} 7 & 6 & 12 \end{pmatrix} \text{ --- a row-matrix}$$

$$B = \begin{pmatrix} 16 \\ 8 \\ 5 \end{pmatrix} \text{ --- a column-matrix}$$

$$\therefore AB = \begin{pmatrix} 7.16+6.8+12.5 \end{pmatrix}$$

$$= \begin{pmatrix} 112+48+60 \end{pmatrix}$$

$$= \begin{pmatrix} 220 \end{pmatrix}$$

The cost of the order is £220

Figure 17.14   Array Multiplication

17.10 EXERCISES

Using these arrays:

$$A = \begin{bmatrix} 2 & 7 & 3 \\ 8 & 1 & 6 \end{bmatrix} \quad B = \begin{bmatrix} 4 & 0 & 5 \\ 1 & 3 & 7 \end{bmatrix} \quad \text{and } C = \begin{bmatrix} 6 & -1 \\ 2 & 5 \\ 8 & 4 \end{bmatrix}$$

calculate each of the following:

(a)   A + B

(b)   3A

(c)   2A – B

(d)   2B – A

(e)   AC

(f)   CA

18 The Application of Arrays

18.1 PRODUCT PRICING APPLICATIONS

In the previous chapter an illustration was given in which two arrays were multiplied together to produce the total cost of a product order. In a straightforward extension of this in the first example, shown in Figure 18.1, two matrices S and P are multiplied together; each of the four rows of S contains details of the order placed by a customer for each of five products. The column-matrix P defines (in £) the price of each of these five products. When SP is evaluated the result in the cost-matrix C which gives the cost of the order for each of the four customers, so that customer 1 has placed an order costing £195 and so on.

$$\begin{pmatrix} 7 & 5 & 4 & 2 & 8 \\ 4 & 3 & 7 & 12 & 2 \\ 1 & 0 & 4 & 8 & 3 \\ 3 & 8 & 19 & 0 & 2 \end{pmatrix} \begin{pmatrix} 8 \\ 13 \\ 2 \\ 9 \\ 6 \end{pmatrix} = \begin{pmatrix} 195 \\ 205 \\ 106 \\ 178 \end{pmatrix}$$
$$\quad\quad\quad S \quad\quad\quad\quad P \ = \ C$$

Figure 18.1 Matrix Multiplication – Product Prices

The extension of this situation leads us to deal with a new matrix P_1 in which there are now three columns, each relating to the prices of each of the five products but at each of three different shops; hence item 1 costs £8 each at shop 1, £11 each at shop 2 and

£6 each at shop 3. Hence the product SP_1 now produces a 4×3 matrix C_1 which gives the cost of each of the four customers orders as it would be at each of the three shops; the first column gives the totals at shop 1, the second at shop 2 and so on. We can see that all customers get the best deals at shop 2; however when comparing shops 1 and 3, the former is more expensive than the latter in the first three cases but the position is reversed for customer 4 (Figure 18.2).

$$
\begin{pmatrix} 7 & 5 & 4 & 2 & 8 \\ 4 & 3 & 7 & 12 & 2 \\ 1 & 0 & 4 & 8 & 3 \\ 3 & 8 & 19 & 0 & 2 \end{pmatrix}
\begin{pmatrix} 8 & 11 & 6 \\ 13 & 8 & 12 \\ 2 & 3 & 5 \\ 9 & 5 & 7 \\ 6 & 4 & 5 \end{pmatrix}
\begin{pmatrix} 195 & 171 & 176 \\ 205 & 157 & 189 \\ 106 & 75 & 97 \\ 178 & 162 & 219 \end{pmatrix}
$$

$$ S \qquad\qquad P_1 \;=\; C_1 $$

Figure 18.2   Matrix Multiplication – Prices for Three Shops

As prices are frequently held in tabular form for a small choice of product lines, the approaches used here are quite common. Note once again that the order used for the matrix multiplication was critical; neither PS nor $P_1 S$ can be defined.

18.2   STACKS

A stack is a data structure found quite commonly in computing applications. In systems software a stack is generally used to handle, for example, the stages of a calculation or for handling processor interrupts; they occur in translation programs, often when evaluating expressions or unscrambling some source code, as well as when passing information from a main program to a subprogram and vice versa.

Physically a stack is a LIFO structure (Last-In First-Out) and it can be likened to a pile of plates; clean plates are generally added to the top of the pile and plates are taken off as required, also from the top of the pile. We do not in general take plates from the bottom of the pile because of the obvious physical difficulties and for the same reason we do not add plates to the bottom either.

The simplest way to represent a stack is by reference to a one-dimensional array (or list) which has two pointers, the first to point to the base of the stack and the second, called the stack pointer, will define the first available space in the stack (Figure 18.3).

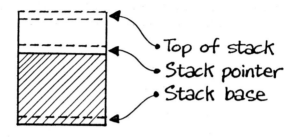

Figure 18.3 A Stack

When a stack is represented by the use of a list in this way the base-of-the-stack pointer will be fixed so as to point to the first element in the list, whilst the stack pointer will be free to move up or down (imagining the stack to be fixed vertically) as items are added to the top of the stack (a process called pushing or stacking) or taken from the top of the stack for use (this process is called popping).

The coding given in Figure 18.4 shows how a stack can be pushed; it assumes that the stack is capable of holding 50 items, although this varies from stack to stack. Hence the base of the stack is held at S(1) and S(50) defines the top of the stack. Before an item such as E can be added to the stack the stack pointer P must be checked to make sure that space does in fact exist in the stack and if not then it must precipitate appropriate action. If space does exist then the item may be positioned in the stack and the pointer adjusted accordingly.

Coding for the purpose of popping the stack can be created easily in a similar manner.

18.3 QUEUES

The queue is yet another data structure; unlike the stack it has a

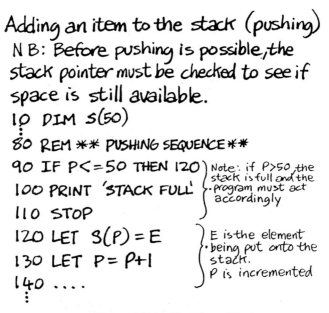

Figure 18.4   Pushing a Stack

FIFO (First-In First-Out) structure, which makes it behave in a manner closely akin to that of a shop queue; the first to join the queue is also the first to leave it. This means that new items are always added to the tail of the queue whereas items always leave from the head. In a computing context the queue is frequently found used for real-time processing, for job scheduling in cases of multiprogramming and for other applications such as data transfer.

As in the case of a stack it is possible to represent a queue by means of a one-dimensional array or list. Again, two pointers are required, the first to point to the front of the queue (the head pointer) and the second to indicate the next available space at the tail-end of the queue (the tail pointer). As items enter and leave the queue so the queue slithers towards the end of the array; to prevent the tail pointer going beyond the end of the space made available for the queue in the array the device of making the tail 'wrap round' to join the queue at the other end is used so that it appears in fact to be chasing the head. Both simple and wrapped-

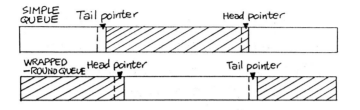

Figure 18.5   Queues

round queues are seen in Figure 18.5, with shading used to denote the queue itself.

Note that at least one open space must always exist between the tail of the queue and the front so that one is able to distinguish between a queue which is empty and one which is full.

In the Basic coding which appears in Figure 18.6 the removal of

Removing an item from the Queue
NB: Check first to see if the queue is empty.
```
10    DIM  Q (50)
  :
200   REM ** REMOVAL SEQUENCE **
205  IF  H <> T  THEN  220
210  PRINT  'QUEUE EMPTY'
215  STOP
220  LET  E = Q (H)
225  IF  H = 1  THEN  240
230  LET H = H - 1
235  GO  TO  245
240  LET  H = 50
245  ...... 
  :
```

Figure 18.6   Removing an Item from a Queue

an item from the head of the queue is undertaken, noting that H and T represent the positions of the head and tail pointers respectively, E being the item withdrawn from the front of the queue. If H = T then the queue is empty but if H = 1 (given that H is not equal to T) then this is a wrapped-round queue and so the head pointer has to be moved to the other end of the array. It is equally straightforward to add items to the tail-end of the queue when required.

18.4   TREES

This particular data structure is used especially in system software in translation programs. To explain the way that a tree is constructed consider the diagram given in Figure 18.7 which shows how a tree structure is capable of representing an ordered set of data, be it numeric or (as in the example) textual. The first item or *datum* (ie the text word 'tree') is stored at the top in the first *node* of the tree, and this is subsequently referred to as the *root node*; to the left and to the right of this datum are the left and right pointers respectively. Since the second datum (the text word 'diagrams') is alphabetically before the first, it is placed at a node on the left descendant from the root node; to position the third datum (the text word 'are') we start from the root node and go down the left descendant since 'are' comes before 'tree' and then down the next left descendant again from the node containing 'diagrams' since it occurs alphabetically before 'diagrams'. In this manner all the datum values get stored, always starting at the root node and going down left or right descendants (according to alphabetic sequence in this case) until they eventually get located at a node. Any node which has no descendants going down from it is called a *terminal node*.

Having now established what a tree structure looks like and how it is created the next stage is to determine how it can be held as an array (Figure 18.8). Each datum, in the order in which it is read in ('tree' is the first datum, so at node 1, 'diagrams' is the second datum so at node 2, etc) has three values associated with it. These values are its actual value (the datum itself), its left descendant (the number of the node to which it descends to the left), and its right descendant. Each node corresponds to the row of the same number in the 9 × 3 array T in Figure 18.8 which holds this informa-

A tree structure to store the
words of the text 'TREE
DIAGRAMS ARE QUITE EASY
WHEN YOU KNOW HOW'

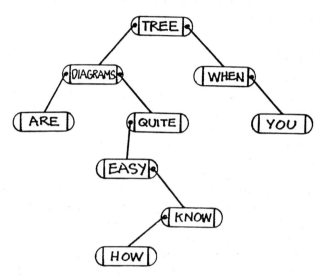

Figure 18.7   Tree Diagram

2	TREE	6
3	DIAGRAMS	4
-1	ARE	-1
5	QUITE	-1
-1	EASY	8
-1	WHEN	7
-1	YOU	-1
9	KNOW	-1
-1	HOW	-1

$T(4,1) = 5$
etc
Tree held as
9×3 array
T

Figure 18.8   Tree Hold as 9 × 3 Array

tion. Each row holds the left descendant, the datum and the right descendant, using −1 in any case where no descendant (left or right) exists.

From the diagram it can be seen that node 4 has a datum of 'quite', it descends on the left to node 5 (which holds datum 'easy') and that there is no right descendant; this can easily be checked now against the original tree structure.

However, a rather more sophisticated 9 × 5 array A can be created to hold the same items in columns 1 to 3 but with the back pointer in column 4 and the trace pointer in column 5 (Figure 18.9). The back pointer gives the number of the node *from* which the current node has descended (or its *parent node* as it is sometimes called). The trace pointer identifies the node whose datum would follow the current datum if all the data was to be arranged in alphabetic order. To illustrate this, consider row 6 (for node number 6) in the next diagram. Since A(6,1) = −1 there is no left descendant, since A(6,2) = 7 the right descendant is to node 7 (whose datum is 'you') and since A(6,3) = 'when' this is the current datum. At this point check back to the tree diagram itself to verify the accuracy of these three values. Now, since A(6,4) = 1 it follows that the parent node is number 1, the root node itself, so that 'when' is directed directly from 'tree'; again this can easily be checked in the diagram. Finally, since A(6,5) = 7 it follows that the datum at node 7, which is 'you', comes immediately after the current datum, 'when' in the alphabet if all the 9 data were arranged in order; again, check the validity of this. Try next the same investigation with reference to other nodes such as numbers 2 and 8. Note that the back pointer for node 1 is −1 since it *is* the root node and so has no parent node; likewise node 7 has a trace pointer of −1 since there is nothing after the datum when sequenced alphabetically.

This revised version of the array takes up more space but it does provide additional facilities. Tree processing involves the insertion or the deletion of data, including therefore the subsequent adjustment of all relevant pointers and descendants as well as searching for specific datum values; programming such processes provides an interesting exercise. It is worth noting too that many database management systems make use of tree structures to assist

2	6	TREE	−1	6	
3	4	DIAGRAMS	1	5	
−1	−1	ARE	2	2	Tree
5	−1	QUITE	2	1	held
−1	8	EASY	4	9	as
−1	7	WHEN	1	7	9×5
−1	−1	YOU	6	−1	array
9	−1	KNOW	5	4	A
−1	−1	HOW	8	8	

LEFT RIGHT   DATUM   BACK TRACE
POINTER POINTER   POINTER POINTER

Figure 18.9   Tree Held as a 9 × 5 Array

in the process of maintaining the information stored in the database itself.

18.5   EXERCISES

1. Produce a tree diagram to hold the following piece of text:

    "Accounting packages include not only the software developed for users".

2. Produce a 10 × 3 array to contain the information about the tree diagram developed in question 1, using left descendant, datum and right descendant for the entries in the three columns.

3. Develop a 10 × 5 array to contain the information about the tree diagram developed in question 1 so that the five columns represent left descendant, right descendant, datum, back pointer and trace pointer.

18.6   TABLE HANDLING

Techniques necessary for the processing of tables and other arrays are more properly dealt with in Chapter 5 of the accompanying book in this series *Programming Techniques and Practice* by A

Chantler so here we shall concentrate far more upon the need for such processing, rather than the ways in which it is carried out.

Having initially created a table one of the first requirements must be the ability to search the table to extract information from it. This is not a very demanding task from a programming point of view; in the case of a two-dimensional array it consists of nothing more than examining every element in the array, possibly on a column-by-column basis, until the element sought is found and then taking appropriate action.

In some cases it may prove better to use a binary chop algorithm if the data items exist in some sequential structure. However, these are simply techniques for finding an element; what is at least as important, if not more so, is to know what is being sought and why, and to understand the structure of the table being searched and what information it actually carries. The 5 column array used for the tree diagram gives some idea regarding how much information can be stored in an array. Consider the problem of searching for all the terminal nodes; these are characterised by both left and right descendants being stored as -1 so the process of finding them will be characterised by statements of the form:

$$\text{IF } A(X,1) = -1 \text{ AND } A(X,2) = -1 \text{ THEN} \ldots \ldots$$

which will successfully carry out the action specified if row X contains a terminal node and not otherwise.

Sorting a table is another standard process which frequently has to be applied. In the case of the tree diagram array we may need to be able to list all the elements contained in datum order, alphabetically or numerically; this can be done solely by reference to the row numbers and the trace pointers. Try to do it!

In the case of a queue holding details of a number of jobs being run in a multiprogramming environment, the queue elements may need to be sorted into a priority order, so implying that a priority level is stored along with the details of each job in the queue.

Multiplying and adding arrays may be achieved very simply in some languages which possess matrix facilities, so that the product of matrix A and matrix B to produce a matrix C may be possible merely by a statement of the form:

$$MAT\ C = MAT\ A\ ^*\ MAT\ B$$

Such a facility would automatically validate A and B to ensure that they can be multiplied at all and would produce an appropriate error message if this were not possible. Other similar statements can be used for addition and subtraction and also for scalar multiplication. In languages which do not have these facilities it is always possible to program round the difficulties to achieve what is needed.

Other table handling routines include merging where, for example, two or more lists may need combining to form one single list. The way that this is done largely depends upon the structure of the lists but again it represents a reasonably straightforward programming task.

Although the section dealing with Product Pricing Applications (18.1) indicated that prices were not always held in arrays, it is worth considering the commercial reality of the situation. If dealing with only a small number of product lines then to house the prices and descriptions in a table held in main memory does provide the fastest possible means of accessing such information and therefore of making use of it. However, if we were to consider a point of sales (POS) application in a supermarket with many tens of thousands of product lines, it is more usual for the data to be held on disk with high-speed disk access facilities to minimise time delays. In these cases data has to be held centrally to service the needs of a large number of checkout points as prices and a list of available product lines may have to be updated on a daily basis, so dynamic is the rate of change in the retail business. Under such circumstances to hold the data totally in a table becomes far too demanding of available processor space, although some systems do compromise between the use of tables and disk access. Nevertheless low-volume applications such as the storage and use of discount rates etc, do make extensive use of tables.

19 Sources of Numerical Error in Data

19.1 INTRODUCTION

A large number of people believe in the wholly fallacious idea that each and every value which emerges as an item of computer output is 100% accurate. Unfortunately, despite all the hype which surrounds computer sales, this promise is less than true; there are a wide variety of totally valid and often substantial reasons why this should be so but the first stage is to acquaint users with reality. However, this is not to say that *all* output is inaccurate and contains not a shred of truth, but the situation needs to be seen in perspective. Errors do occur for reasons, many of which can be controlled or reduced, but rarely can they be totally eradicated; generally, if there are errors at the output stage they are very minor and can be ignored completely. In commercial work, because of the way in which data is held, there may be fewer errors than in more scientific applications, but occasionally (and fortunately rarely) a situation is encountered in which errors get totally out of proportion and are capable of swamping the 'true' value.

Many of the sources of error can be controlled, even if they cannot be completely removed. This chapter attempts to identify many of the sources of error, whilst Chapter 20 will deal with the ways of controlling them; because certain of the methods may involve quite mathematical approaches, and are beyond the scope of this text, such treatment will be kept as simple as possible to ensure that at least the concepts are understood.

19.2 TRANSCRIPTION ERRORS

On at least two occasions data is likely to be copied from one medium to another and in both cases human beings are involved in such copying. These occasions are:

(i)   at the creation of source data in a user department;

(ii) at the creation of machine-readable form (data preparation stage).

Because we are human, even the most accurate among us is likely to occasionally miscopy, especially when working with numbers and when working under pressure, so that even if only one mistake is made for every 1000 digits we copy (and that is exceptionally accurate in practice) then it can be appreciated that in the many millions of digits which exist in data files, just as many thousands of mistakes have been made.

Several identifiable forms of transcription error can be defined:

(a)   Single transposition, where one digit changes places with another,
        eg 237426 becomes 273426.

(b)   Double transposition, where three digits all 'move round' so that two single transpositions have actually occurred,
        eg 963847 becomes 938647.

(c)   Inclusion of an extra digit,
        eg 21356 becomes 213456.

(d)   Loss of a digit,
        eg 9476305 becomes 946305.

(e)   Repeating a digit (which is a special case of (c)),
        eg 274889 becomes 2748889.

(f)   Other random cases,
        eg 28743 becomes 28643,
        3824 becomes 37624,
        10256 becomes 10258.

19.3   SOURCE ERRORS

There are other mistakes that can be made in the user department; these include *actual* errors as distinct from copying errors, for example, by putting down the wrong price for a product as distinct from miscopying the correct price, or claiming to have worked for more hours in a particular week than have actually been worked. These mistakes (some of which may even be deliberate) can often only be prevented by the design of the system as a whole by ensuring, for example, that all source documents are checked and countersigned by a supervisor when they relate to the hours that an employee actually works or by establishing a computerised control system with some sort of clocking on/clocking off routines so that hours are machine logged. System design has a great deal to do with the process of establishing a series of controls to prevent some of the potential for abuse within a system.

19.4   FINITE COMPUTER WORD LENGTH – ERRORS

In Chapters 3 and 4 the use of the computer word as a unit of storage was introduced; at the same time some of the problems caused by the storage of fractional values were also mentioned. Some of these problems are now examined in rather greater detail.

19.4.1   Truncation Errors

Truncation, or rounding down, involves the removal of any digits at the least significant end of a number to enable it to fit into the storage space allocated to it. As these digits are nearly always at the less significant end of the fractional part of the number, it means that the number as it actually gets stored is always less than, or sometimes equal to, the number before truncation took place. If this were to happen with a solitary, one-off, data value then it is unlikely to cause any serious problems. If however, a large number of different truncations were involved during various stages of a calculation then it is possible that the effect might become cumulative and so cause serious errors.

The effect of truncation may be either to decrease or to increase the value of a calculation depending upon whether the quantity being truncated was in the numerator or the denominator of a fraction or whether it was being added or subtracted. To illustrate the effect consider the calculation of the expression $\dfrac{a + b}{c - d}$ in which $a = 3.62841$, $b = 5.38634$, $c = 8.32174$ and $d = 8.31079$ then the accurate result will be the result of dividing 9.01475 by 0.01095 which gives a value of 823.26484. However, if the values of a, b, c and d were each truncated to 3 decimal places because of the limitations of storage then they would become $a = 3.628$, $b = 5.386$, $c = 8.321$ and $d = 8.310$; this time the result of calculating $\dfrac{a + b}{c - d}$ would be the result of dividing 9.014 by 0.011 which produces a solution of 819.45455. This second result is 3.8102946 less than the exact value which indicates an error of about $1/2\%$. The individual percentage errors in a, b, c and d are, respectively, 0.011%, 0.006%, 0.009% and 0.009% so that the effect of these relatively small changes, which total some 0.035% is to produce a result with an error of about 14 times as large.

19.4.2 Rounding Errors

Rounding a number *up* will always tend to increase its value just as rounding *down* (or truncation) will likewise tend to reduce its value. The circumstances and the effects are closely akin to those of truncation in the previous section. In general, however, we round *off*, so that a number will be rounded *up* if this causes the least change in its value, but rounded *down* if this is what is needed to cause the minimum adjustment to its value. The 'half-way house' situation causes 7.325 to be rounded off to 7.32 (to 2 decimal places) but causes 7.335 to be rounded off to 7.34; this is a fairer system than always rounding 5 up as is still sometimes done. The rule here is that if the figure which is to be the last one kept after rounding (in this case the second place of decimals) is even before rounding then the 5 is rounded *down* whereas if it is odd before rounding then the 5 is rounded *up;* in either case the figure will be even afterwards. Rounding off to so many decimal places or to so many significant figures are both specific cases of rounding.

19.4.3 Overflow Errors

Truncation and rounding errors have both been concerned with loss of accuracy at the least significant (right-hand) end of a number. However it has already been explained that a computer word also has a maximum capacity which depends upon the number of bits it contains. If a certain amount of space were allocated for a numeric value and the value is too large to go into it then truncation occurs at the most significant (left-hand) end; for example, if we declare a field in a Cobol program to have a *picture* of 999 and attempt to store the value 2734 in it, the result will be to truncate the 2 and to leave a value equal to 734 stored instead. Such errors are clearly very serious and control over them is most vital.

19.5 CONVERSION ERRORS

Since numbers are stored inside the computer in binary, it follows that when decimal fractions are converted upon input to binary fractions there is the distinct possibility that the conversion will not be exact and as a result, some truncation will occur. The truncated value may then be used in calculation, leading to a series of further 'knock-on' errors and when output, will not have the same decimal value that was input. Thus if we wanted to store the decimal fraction 0.3 (which has a bicimal value of 0.010011001 . . .) in 6 bits it will be held as 0.010011 which is the decimal value 0.296875; so a conversion error has occurred.

19.6 ERRORS IN ALGORITHMS

A calculation may sometimes be undertaken by two or more different methods, each essentially correct, yet one method may produce a more accurate result than the other(s). Such a method is said to have a smaller algorithmic error; the usual reason for it being more accurate is that it involves a smaller number of stages and since any rounding or truncation errors tend to build up at each stage, the process which has the smallest number of stages is the one most likely to have the least build-up of errors.

19.7 FLOATING-POINT FORMAT ERRORS

We generally tend to associate errors with fractional values, except for overflow situations. However, even integral values can contain sources of error if held in floating-point format. Consider (using decimal values) the storage of the integral value 27594 in a floating-point format in which only four places have been allocated for the mantissa. In such a case the value would be stored as $+0.2759 \times 10^5$ giving an immediate error, since 27594 is being held as though it were 27590.

19.8   RELATIVE AND ABSOLUTE ERRORS

Absolute error is defined to be the difference between the true value of a number and the value as used to represent that number. In an earlier example in this chapter on conversion errors, we saw that 0.3 was held in binary as though it were really 0.296875, so that in this case the absolute error is:

$$\text{value as stored} - \text{true value}$$
$$= 0.296875 - 0.3$$
$$= -0.003125.$$

Having provided the definition for absolute error relative error is defined to be:

$$\text{absolute error/true value}$$

so that, using the same values we now have a relative error of:

$$-0.003125/0.3$$
$$= -0.0104166 \text{ (or } -1.04166\%).$$

Solve $105x + 100y = 305$
$$Kx + 99y = 303$$
where $K = 105 \pm 5$
ie the possible error in K is 4.76%
If K takes its central value 105,
then the solution is $x = 1$ $y = 2$
BUT if $K = 100$ (ie its lower limit)
then we have: $105x + 100y = 305$
$$100x + 99y = 303$$
whence $x = -0.2658...$
$$y = 3.329...$$
which means that a 4.76% error in K
has led to an error of 126.58%
in x and an error of 66.45%
in y

Figure 19.1   Ill-conditioning

19.9   ILL-CONDITIONING

This phrase refers to a situation in which small errors which exist in
the values of the data items used in an expression cause larger
errors in the calculation of the expression itself; such a case was
seen in the early part of this chapter when $\dfrac{a + b}{c - d}$ was calculated.
The extent of this type of error is in fact a good deal greater than it
is usually given credit for.

In Figure 19.1 an attempt is made to solve a pair of simultaneous
equations; one of the quantities involved, K, is liable to an error of
at most 4.76%, but the results show that this can cause errors of
well over 100% in the value of x.

20 Control of Errors

20.1 TRANSCRIPTION ERRORS AND VERIFICATION

The previous chapter looked at the ways in which errors could occur in source documents. Now some at least of the means whereby we can control the extent of these errors must be investigated. Whilst quite a number of these steps are aspects of a total systems design, many of them are quite straightforward and easy both to understand and to implement.

When data is sent from a user department to the computer room it is usually on source documents; these do have a tendency to get mislaid, so before leaving the user department it is normal practice to fill out a batch control slip which then accompanies each batch of source documents. On this control slip may be one or two totals, the first possibly to indicate how many documents are included in the batch, whilst the second is likely to be the result of totalling a selected field on each document for the whole batch. These totals enable an automatic check to be made upon receipt at the computer department to ensure that if any documents do get mislaid, then at least their loss will be notified; the field chosen for totalling may provide a sensible batch total if it represents the number of hours an employee has worked, or a nonsense or hash total if it involves adding together fields such as account numbers whose total clearly is otherwise meaningless.

As soon as the process of creating machine-readable data has been started in the data preparation area, verification may be carried out to reduce the risk of errors in transcription. When all the records in a batch have been entered, they are checked in one of two ways, preferably by a second key-punch operator, who is not so likely to repeat any mistakes made by the person who first keyed in the data. The first way involves visually checking each record as it is brought up on the screen against the source document looking for errors; the second involves re-typing the data from the same source documents but with the machine set to 'verify' mode. When the keys are pressed whilst the machine is in 'verify' mode they are compared against the data as it is already held to check that they are indeed the same; if they are not then the machine will respond in some way to enable correction to be made before continuing. In this way the majority of transcription errors can be eliminated 'off-line' before the file is submitted for live processing. There is no absolute guarantee that *all* errors *will* be found since both operators might mis-read the source document in the same way or the error may originate back in the user department which means that neither key-punch operator is likely to detect it.

Since most of these errors are 'human' it is very difficult to do better than to eliminate the majority of them. Frustrating though it is, this is one area of computer accuracy in which we are able to exercise less than perfect control. It is important therefore not only to carry out verification as described but also to design the whole system as carefully as possible in an attempt to minimise the number of errors that can occur.

20.2   VALIDATION TECHNIQUES

The previous section dealt with checks that could be made in order to reduce errors in an off-line environment. Once data has been loaded into a machine-readable format, on disk or tape in particular, the computer-controlled validation process can be used to check for errors on-line. At this stage it is acknowledged that there may still be (indeed almost certainly *will* be) errors in the records within the files and that certain of these could not have been detected by verification; if a code number were entered wrongly on the source document then the key-punch operator can hardly be blamed for transcribing it as seen.

Quite commonly, in batch situations, there will be a full program written to carry out the tasks of validation, a 'data vet' program which accepts as input the unchecked source data and outputs a file of accurately checked data which is error-free plus details of all those records not output to that file because they contain definable errors. This gives staff the opportunity to correct any errors reported and ultimately to ensure that all records submitted have undergone the validation scrutiny and that all have had any errors put right. The diagram in Figure 20.1 is typical of a data vet carried out as part of a batch system.

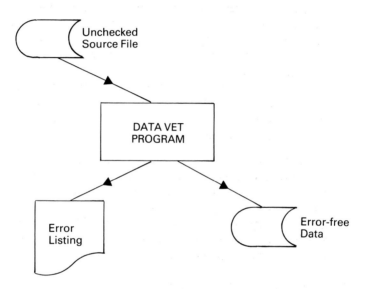

Figure 20.1   Data Vet as Part of a Batch System

In the situation in which real-time processing is to take place, data validation has to be done with each record individually as it is received, causing erroneous data to be turned round at once to the user before processing can take place.

Validation consists of establishing a series of criteria, which will vary for different situations, and which must be met by some or all of the fields contained in a record or fields calculated on the basis of such data. Validation is used to establish 'reasonableness' and ensures that data which is to be processed later satisfies a number of criteria such as:

(a) A quantity field must be numeric (consisting only of digits 0 to 9 etc).

(b) A numeric field must lie within certain 'acceptable' limits, for example a 'number of hours worked in a week' may be required to lie between 35 and 65 inclusive.

(c) A field may be required to contain a specified number of characters, for example an account number may be exactly 6 digits long.

(d) A field may be required to contain certain specific characters if it is valid.

(e) A field may be checked to make sure that it is actually present.

(f) A reference number may have to satisfy check-digit rules (detailed in full in section 20.3).

Examples of these tests appear in Figure 20.3 on page 306.

Once again, just because an item of data *does* satisfy all the validation tests imposed, it does not guarantee its accuracy. If a 'number of hours worked' is recorded as 47 in the user department instead of as 41, then neither verification nor validation can possibly detect the error if both 47 and 41 are feasible; this is a case in which the only real check which can succeed lies in the user department and will form an important aspect of systems design.

20.3   CHECK DIGITS – A VALIDATION DEVICE

If we create a six-digit number, for example, as a customer account reference number, then an attempt must clearly be made to guard against the possibility of it being quoted in error, lest the wrong customer is invoiced for goods ordered. One method of doing this is to add a seventh digit, a check digit, according to some agreed algorithm. The method described here is not unique but the approach used is typical. To each of the six digits is allocated a 'weight', being 7 for the first, 6 for the second and so on down to 2 for the sixth, on a left to right basis. Next, each digit is multiplied by its weight and the resulting products are added together to form a total. This total in turn is divided by 11 (which has been chosen as the 'modulus' for the system) and the remainder found as a result of the division is then subtracted from the same modulus, 11, to yield the check digit which is now added, as the seventh digit, to the account reference number.

If the number had only possessed four digits instead of six, then weights of 5, 4, 3 and 2 respectively would have been added and a fifth digit calculated as the check digit using the same general approach. Any length of number can be catered for by a similar method with no great difficulty. Although 11 was chosen as the modulus in the example in Figure 20.2, there are other possibilities

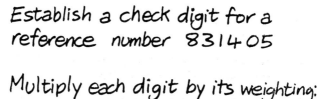

Establish a check digit for a
reference number 831405

Multiply each digit by its weighting:
8 × 7 = 56
3 × 6 = 18
1 × 5 = 5
4 × 4 = 16
0 × 3 = 0
5 × 2 = 10
Total of products 105

Divide total by modulus
105 ÷ 11 = 9 remainder 6
Subtract remainder from modulus
11 - 6 = 5. Check digit is 5
∴ New reference number is 8314055

Figure 20.2   Example of Working out a Check Digit

which could have been chosen such as 13, 17 and 19 since they are prime numbers and all being of size 11 or greater. However, using 11 as the modulus it is quite possible that the check digit might turn out to be 11 or 10; in such cases it would generally be quoted as O or X to ensure that a single 'digit' appears. Other equivalent systems do exist but the principles of weights and a modulus apply in each of them.

Having once established a reference number *with* a check digit it is an easy matter to verify that any 'stated' reference number matches up to the rules given and that the check digit itself is correct for the combination of the other digits present; this helps to reduce enormously the problems created by transposition of digits when recording such things as a customer reference number.

In the diagram given in Figure 20.3 are examples of a variety of

VALIDATION EXAMPLES

① Check digits (using modulus 11 and weightings of 5, 4, 3, 2)
The 5-digit number 27103 (in which 3 is the supposed check digit) is valid
The 5-digit number 86043 is not valid

② 39246 is numeric but 39Z64 is not

③ For a field supposed to have 6 digits
290156 is valid whereas 29156 is not

④ If a rate of pay must lie between 5.50 and 12.75 inclusive
then 6.38 is valid
but 4.32 is not valid
nor is 13.65

⑤ If a stock number must begin with the letter A or the letter B and then have a 3 digit numeric part,

A 805 is valid as is B647
but C926 is not valid
nor is A12

Figure 20.3   Examples of Validation Criteria

different validation criteria in use, including one using the check digit criterion.

20.4 ROUNDING ERRORS

Although these can never be completely eradicated their effect can be restricted by taking the following precautions:

(a) Never truncate or round up exclusively if at all possible, since both distort results over a period of time; instead always round off. This will lead to unbiased rounding and to ensuring that values are so stored that they are neither consistently low (as the result of truncation) nor consistently high (the consequence of rounding-up all the time) but that all the component parts of an answer tend to balance out with, in effect, as many truncations as there are roundings-up. With binary quantities rounding-off is even easier than with decimal values; in order to round off to 5 bicimal places round down if the sixth place is occupied by a 0 and round up if it is occupied by a 1.

(b) In floating-point arithmetic the same rule with regard to rounding must be made to apply. Note that in such cases errors may be magnified if they impinge upon the value of the exponent in particular.

(c) Conversion errors can be minimised in the same way.

(d) Always use the maximum space available for the holding of intermediate results so that absolute errors are kept as small as possible and to avoid the effects of cumulative build-up. This applies even in cases where the final result has only to be quoted to two places of decimals.

20.5 ORDER OF OPERATIONS

In order to further reduce the effects of rounding errors it is always best to add together (or to subtract) a set of values in ascending order of their magnitude. In order to illustrate this Figure 20.4 shows how four values of very different order of magnitude can be added together in three different ways. The first way is exact, in which case the order is irrelevant. The second involves using the

Add together 0.003574 + 0.002807
      + 0.019276 + 0.154308

① Exact value is 0.179965

② Add together in descending order:

(a) 0.154308 + 0.019276 = 0.173584

   Round to 3sf   = 0.174000

(b) 0.174000 + 0.003574 = 0.177574

   Round to 3sf   = 0.178000

(c) 0.178000 + 0.002807 = 0.180807

   Round to 3sf   = 0.181000

③ Add together in ascending order:

(a) 0.002807 + 0.003574 = 0.006381

   Round to 3sf   = 0.006380

(b) 0.006380 + 0.019276 = 0.025656

   Round to 3sf   = 0.025700

(c) 0.025700 + 0.154308 = 0.180008

   Round to 3sf   = 0.180000

Figure 20.4   Order of Operations – Rounding Techniques

four quantities in descending order of magnitude, with rounding at each stage to three significant figures; the third method involves the use of the same level of rounding but this time with the quantities handled in ascending sequence.

When rounded to three significant figures it is at once apparent that the third result is closer to the true (final) result but it should not be assumed that the 'descending order of magnitude' approach is always going to be the worst to use. However, it is quite certain that the 'ascending order of magnitude' approach is always the safer to adopt consistently whenever one is able to exercise control over the order in which the calculations are to be carried out.

20.6   CORRECTION OF ALGORITHMIC ERRORS

In the previous chapter it was stated that the choice of algorithm for the solution of a problem often influenced the accuracy of the result, even if each algorithm used was legitimate for the solution of the particular problem under consideration. If the algorithm happens to involve the calculation of fractions or of other values in which rounding errors may occur (in floating-point arithmetic etc) then it is possible that these errors may build up and lead to an unacceptable level of inaccuracy in the result obtained or even in the very control of the algorithm itself.

Consider the case of a program loop which is controlled by a counter X which is to start off from the value 0 and to go up to 8 inclusive, in steps of decimal 0.1. The arithmetic is to be performed using 6 bicimal places with rounding at each stage to correct the value to 5 places. The 6-bit binary value for decimal 0.1 is 0.000110 and this already produces a sizeable error since this equates exactly to decimal 0.09375 (rather than to 0.1) so the rounded value to 5 places is therefore 0.00011.

Looking at the first two columns in Figure 20.5 we can see the effect upon X of these errors at each step; by step 10 X ought to be exactly 1 (since it should be 0 + 10 × 0.1) but in fact it is rather less. By step 20 it should be exactly 2 but it equates in fact with decimal 1.875 and so if the values of X are to be used for calculations within the loop then there are likely to be considerable errors as a direct result.

If, however, the loop was controlled by the use of the integer counter C which goes up in steps of 1 from 0 to 80 inclusive, but allowing X to be calculated as C/10 at each stage, then the third column contains the values of X so obtained; whilst they still contain some error, by step 10 the value has been corrected so that it is exact as it will be again once step 20 has been reached. Hence, although the intermediate values may indeed not be exact, they certainly do produce far more accurate evaluations from step 10 onwards. Whereas the error in the first method gets progressively worse in absolute terms, the error in the second never exceeds the worst case which is experienced at step 9.

One lesson to be learned from this is clearly to make use of as

X − values

STEP	FIRST METHOD	SECOND METHOD
1	0.00011	0.00011
2	0.00110	0.00110
3	0.01001	0.01001
4	0.01100	0.01100
5	0.01111	0.01111
6	0.10010	0.10010
7	0.10101	0.10101
8	0.11000	0.11000
9	0.11011	0.11011
10	0.11110	1.00000
11	1.00001	1.00011
12	1.00100	1.00110
13	1.00111	1.01001
14	1.01010	1.01100
15	1.01101	1.01111
16	1.10000	1.10010
17	1.10011	1.10101
18	1.10110	1.11000
19	1.11001	1.11011
20	1.11100	10.00000
21	1.11111	10.00011
22	10.00010	10.00110

Figure 20.5   Program Loop Showing Cumulative Errors

much integer calculation as possible, converting from integer to fractional value only when it is absolutely necessary to do so.

20.7   NESTING METHODS

If we had to calculate the value of the polynomial expression $4x^3 - 7x^2 + 5x + 19$ for any given value of x, then there would be potentially several different calculations to be undertaken, each of which will contain a source of error if x is non-integral, so the number of rounding errors could lead to a great deal of inaccuracy. This clearly would be compounded still further if the polynomial had to be evaluated for each one of a range of x-values.

In order to compensate for this it is possible to 'nest' the expression as follows:

$$4x^3 - 7x^2 + 5x + 19$$

$$= (4x^2 - 7x + 5) x + 19$$

$$= ((4x - 7) x + 5) x + 19$$

in which a linear expression $4x - 7$ is nested inside another and so on. The calculation is shown in Figure 20.6.

To evaluate $4x^3 - 7x^2 + 5x + 19$
for $x = 3.6$ use $((4x-7)x+5)x+19$
and compute from $4x-7$ outwards:
$((14.4-7)*3.6+5)*3.6+19$
$= (7.4*3.6+5)*3.6+19$
$= (26.64+5)*3.6+19$
$= 31.64*3.6+19$
$= 113.904+19$
$= 132.904$

To evaluate $8x^2+5x-6$
Use $(8x+5)x-6$, etc.

To evaluate $9x^4-7x^3+4x^2+5x-3$
Use $(((9x-7)x+4)x+5)x-3$
etc.

Figure 20.6   Nesting Methods

This approach allows for a reduction in the number of calculations that have to be accomplished at each stage and so will tend to reduce the amount of error that may arise. Even if integral values of x do not lead to error risks the nesting method does provide for a faster means of evaluating such polynomials in all cases. In particular, this method is valuable in cases where the expression has to be calculated within a loop for a large number of different x-values; apart from being less error-prone it is also faster and well-suited to calculator usage even if a computer is not being used.

20.8   CONCLUSIONS

Errors, regrettably, are unavoidable but the purpose of this chapter and Chapter 19 has been to identify some of the areas in which errors are most likely to occur and certain of the methods that can be used to contain them. Human errors will always be present in some measure but good system design, verification and validation offer methods of reducing them if not of eliminating them. Errors caused by virtue of the computer word-size are also unavoidable and even if an environment in which these can be minimised were created they will still be there in some degree; even hand-held electronic calculators suffer from the same drawback. Whenever there is a lot of calculation to be done, it is important to remember the problems created by failure to consider algorithmic design.

Errors caused by transmission failure and other forms of hardware fault are generally insignificant and are usually random in their pattern of occurrence and cannot therefore be planned for.

Appendix 1

Solutions to Exercises

CHAPTER 1

1.2

(a)	120	(h)	−9	(o)	27	(v)	−126
(b)	21	(i)	−8	(p)	20	(w)	35
(c)	−104	(j)	−5	(q)	6	(x)	19
(d)	−102	(k)	10	(r)	5	(y)	8
(e)	−198	(l)	12	(s)	23	(z)	11
(f)	12	(m)	9	(t)	5		
(g)	5	(n)	21	(u)	24		

1.3

1. 74,000
2. 121
3. 18
4. 68
5. 4368
6. (a) 1,362,500 (b) 637,500 (c) 411,422

1.5

1. (a) $^4/_5$ (b) $1^{15}/_{28}$ (c) $5^{17}/_{72}$ (d) $^3/_8$ (e) $3^1/_6$ (f) $^{11}/_{28}$
   (g) $^1/_{12}$ (h) $^9/_{20}$ (i) $1^1/_3$ (j) $1^1/_2$ (k) $^1/_4$ (l) $1^1/_3$
2. 96

3. 24

4. 8

5. (i) 108   (ii) 72   (iii) 5.2 secs

6. (i) 4.5 hrs   (ii) 38,250   41,400   44,100

1.7 1. 2003.167

2. 47.57

3. 44.814

4. 12.9

5. $187.25

6. (i) 7.205 Mbytes   (ii) 2.795 Mbytes

7. (i) 7   (ii) 5

1.9 1. 10

2. 7:3

3. £204

4. 25

5. 36

6. 76 square metres

CHAPTER 2

2.3 1. £90

2. £5940

3. £1581

4. (i) £530   (ii) £561.80   (iii) £595.51

5. £185

6. 19.44%

7. (i) £1,144,000   (ii) £1,006,720

8. (i) 68.25 Kbytes   (ii) 71.6625 Kbytes (or 71663 bytes)

9. $7^1/_2\%$

10. 18.75%

11. 4.17%

12. 22.62%

2.5   1. (a) $4^5 = 1024$ (b) $2^8 = 256$ (c) $5^4 = 625$
(d) $3^7 = 2187$

2. (a) 1   (b) $^1/_9$ (c) $^1/_4$ (d) 2   (e) 5   (f) 2

CHAPTER 3

3.4   1. (i) 6   (ii) 9   (iii) 29   (iv) 13   (v) 19   (vi) 50
(vii) 26   (viii) 115   (ix) 205   (x) 150

2. (i) 15   (ii) 21   (iii) 35   (iv) 110   (v) 132
(vi) 242   (vii) 263   (viii) 459   (ix) 862   (x) 1485

3. (i) 20   (ii) 35   (iii) 27   (iv) 60   (v) 125
(vi) 162   (vii) 2739   (viii) 2821   (ix) 4013
(x) 62,014

4. (i) 1.25   (ii) 6.625   (iii) 0.1875   (iv) 2.296875
(v) 5.109375   (vi) 8.453125   (vii) 3.6875
(viii) 25.14453125

3.6   1. (i) 1101   (ii) 11001   (iii) 11101   (iv) 1000000
(v) 1001001   (vi) 1111111   (vii) 1100111
(viii) 10101111   (ix) 11101100   (x) 11000111
(xi) 100010010   (xii) 111101101

2. (i) 11   (ii) 21   (iii) 27   (iv) 47   (v) 57
(vi) 123   (vii) 140   (viii) 144   (ix) 177   (x) 455
(xi) 1024   (xii) 1436

3. (i) 17   (ii) 1D   (iii) 23   (iv) 2A   (v) 3B

    (vi) 4F   (vii) 61   (viii) 8C   (ix) FE   (x) 104

    (xi) 1FD   (xii) 219   (xiii) 50F   (xiv) 7CE

    (xv) ABC   (xvi) 2F0F

3.10 1. (i) 0.10000111   (ii) 0.11101000   (iii) 0.01001011

   2. 0.310550

   3. 0.33F7C

   4. (i) 556   (ii) 614   (iii) 33   (iv) 26   (v) 0.65

       (vi) 0.54   (vii) 3.2   (viii) 13.54

   5. (i) 111010101   (ii) 101011110   (ii) 10011.1

       (iv) 1000110.101011

   6. (i) D3   (ii) 25   (iii) 0.E   (iv) 5.8

   7. (i) 101011   (ii) 111111000111   (iii) 1000.11

       (iv) 111101.1110001

3.14 (i) 100100   (ii) 110001   (iii) 110010   (iv) 1101000

   (v) 101110   (vi) 1000101   (vii) 11111101   (viii) 10111

   (ix) 100100   (x) 101110

CHAPTER 4

4.3 1. 001011011001   7. −820

   2. 01100001   8. 11000111

   3. 0.00101111   9. −61

   4. 0.0010111101001110   10. −512 to +511

   5. 0010010.11100   11. 11.671875

   6. 111011101111   12. 1011110001

4.5 1. (i) 34 or 00100010_2 (ii) 79 or 01001111_2

2. (i) 257 or 000100000001_2 (ii) 1427 or 010110010011_2

(ii) 674 or 001010100010_2

4.7 1. 000001100001 and 1552

2. 6

3. 001000011001   (i) 67   (ii) 67

4. 010011011011   (i) 38   (ii) 39

4.9 1. 0001 0101 0010 0110

2. 0010 0101 0011
$$+$$
0100 0010 0110
<hr>
0110 0111 1001   and 679
<hr>

3. 0001 0011 0101 0110
$$+$$
0010 0110 0111
<hr>
0001 0110 0010 0011   and 1623
<hr>

4.15 1. 001001010000

2. 111011101100

3. $^1/_{000100111100}$ or decimal 316

4. 0.101111000010

5. 000011000011.011011010100

6. -786

7. 53

8. 000010101100   (i) 5   (ii) 5

9. 000000100101_2 010010100000_2 or 1184

10. 01111001 and 010100110110 giving 011000010101 or 615

CHAPTER 5

5.4 (i) 0.74×10^7;   0.74;   7;   0.74E7

(ii) 0.256×10^{10};   0.256;   10;   0.256E10

(iii) 0.628×10^{-4};   0.628;   −4;   0.628E−4

(iv) 0.153×10^{-9};   0.153;   −9;   0.153E−9

(v) -0.82×10^7;   −0.82;   7;   −0.82E7

(vi) -0.554×10^{11}; −0.554;   11;   −0.554E11

(vii) -0.27×10^{-3};   −0.27;   −3;   −0.27E−3

(viii) -0.79×10^{-7};   −0.79;   −7;   −0.79E−7

5.7 1. (i) 000110111011   (ii) 011111101101   (iii) 100110100111
(iv) 111111100001

2. (i) 010110111011   (ii) 001111101101   (iii) 110110100111
(iv) 101111100001

3. − 127 to −0.000030518 and +0.000030518 to +127

4. −2,143,289,300 to -4.54747×10^{-13} and $+4.54747 \times 10^{-13}$ to
+2,143,289,300

5.10 1. (i) 85   (ii) 255   (iii) 417   (iv) 4   (v) 1

(vi) −2   (vii) −3   (viii) 4   (ix) 0   (x) −3

2. (i) 0.10011×2^5

(ii) 0.10100×2^6

(iii) 0.10100×2^2

3. (i) 0.10100×2^2

(ii) 0.10011×2^7

(iii) 0.10111×2^3

4. (i) 0.1100×2^6

(ii) 0.1100×2^9

5. (i) 0.101×2^3

 (ii) 0.100×2^8

6. $(0.101010 \times 2^6) + (0.100101 \times 2^7)$ giving 0.111010×2^7 or 000111111010

CHAPTER 6

6.4 1. (a) Quantitative, continuous.

 (b) Quantitative, discrete.

 (c) Qualitative.

 (d) Quantitative, continuous.

 (e) Quantitative, discrete.

 (f) Quantitative, discrete.

 (g) Quantitative, discrete (in practice).

 (h) Qualitative.

 (i) Quantitative, discrete.

 (j) Quantitative, continuous.

2.

Salary($)	Tally	No of staff
60–	/ / /	3
80–	/ /	2
100–	/ / / /	4
120–	/	1
140–	++++ / / / /	9
160–	++++ / / / /	9
180–	++++	5
200–	++++ /	6
220–	/	1
240–	/ /	2

260 –	++++ /	6
280 –	++++ ///	8
300 –	////	5
320 –	/	1
340 –	////	4
360 –	///	3
380 – 399	/	1

3.

Time (secs)	Tally	Frequency
275 –	++++ /	6
300 –	++++ /	6
325 –	///	3
350 –	++++	5
375 –	//	2
400 –	++++ ///	8
425 –	//	2
450 –	///	3
475 –	////	4
500 –	////	4
525 –	///	3
550 –	//	2
575 – 599	//	2

6.6 1. Pictogram using human figures probably, all of equal size, best if each is to represent 25 people.

2. (i) Pie chart angles are: IBM 72°, ICL 42°, Unisys 24°, Tandem 30°, Honeywell 54°, Hewlett Packard 42°, DEC 84°, Data General 12°.

   (ii) Heights proportional to frequencies; widths of 'bars' equal; order does not matter.

4. For May 1983 the angles are: Industrial Action 70°, Hardware Faults 81°, Software Faults 174° and Human Error 35°. For May 1987 angles are 15°, 37°, 249° and 59°; radius is 1.26 times that used in 1983.

5. Data is continuous; diagram follows as from section 6.5.4.

6. Data is discrete; diagram follows as from section 6.5.4.

6.10   1.

Age, less than, (yrs)	20	25	30	35	40	45	50	55	60	65
cf	0	12	29	50	68	82	95	107	116	120

2. Initial downwards trend, then rising quite sharply by end of period. Tuesday consistently better than other days.

3. There is evidence of inverse correlation between x and y.

4. There is no clear evidence of any pattern.

CHAPTER 7

7.5   1.   (i) $1/10$ (ii) $7/20$ (iii) $1/4$ (iv) $9/10$ (v) $3/10$

(vi) 1   (vii) $9/10$

2.   (i) $17/30$ (ii) $4/15$ (iii) $2/15$ (iv) $13/30$ (v) $1/6$

(vi) 0

7.7   (i) $4/5$ (ii) $1/5$ (iii) $8/15$ (iv) $3/20$ (v) $1/60$

(vi) $2/5$ (vii) $1/25$ (viii) $4/125$

7.9   1.   (i) $11/30$ (ii) $4/15$ (iii) $3/10$ (iv) $7/15$

(v) $3/5$ (vi) $1/300$

2.   (i) 0.8   (ii) 0.3   (iii) 0.7   (iv) 0.04   (v) 0.1

(vi) 0.3   (vii) 0.2   (viii) 0.2

CHAPTER 8

8.3 1. (a) $14^5/_7$ (b) 27 (to nearest integer).

2. (a) 2.9 sec (to nearest 0.1 sec).

    (b) 2.6 sec (to nearest 0.1 sec).

    (c) 10.3%.

3. 6.3 years (to nearest 0.1 year).

8.5 1. 23   26   28   29   29   32   35   37   38   41   42   49   51

median is 35 years.

2. 28.5 minutes.

8.7 1. 5

2. Beige.

8.9 1.

x	f	$d = x - 19$	fd
17	4	−2	−8
18	8	−1	−8
19	19	0	0
20	12	1	12
21	5	2	10
22	2	3	6
	$\Sigma f = 50$		$\Sigma fd = 12$

Mean $= 19 + {}^{12}/_{50} = 19.24$ calls/day.

2. 76.3 pages/job.

3. Mean is 155 errors/hour (to nearest integer).

4. Mean is $7255.

8.11 1. 5

2.

Salary less than	4000	5000	6000	7000	8000	9000	10,000	11,000	12,000
cf	0	7	26	64	109	143	170	191	200

8.13 1. 95

2. (ii) $35 + 5 \times \dfrac{7}{7+9} = 35 + \dfrac{35}{16} = 37.1875$ or 37.2 years

CHAPTER 9

9.4 1. Range is 29; interquartile range is 12.

2. (i) Increase to 32   (ii) no change at all.

3. 65

4. (i) 35   (ii) 15   (iii) 7.5

5. (i) no change   (ii) 17   (iii) 8.5

9.6 1. Mean is 6.8 minutes.

Deviations are:
−0.7 0.5 −1.4 0.1 1.9 −1.5 −0.4 1.5.

Squared deviations are:
0.49 0.25 1.96 0.01 3.61 2.25 0.16 2.25.

Sum of squared deviations = 10.98.

Mean of squared deviations is 1.3725.

Standard Deviation is 1.1715 minutes.

Variance is 1.3725 minutes².

2.

x	$x-200$	$d = \dfrac{x-200}{50}$	f	fd	fd^2
50	−150	−3	2	−6	18
100	−100	−2	7	−14	28
150	−50	−1	15	−15	15
200	0	0	21	0	0
250	50	1	16	16	16
300	100	2	14	28	56
350	150	3	12	36	108
400	200	4	8	32	128
450	250	5	4	20	100
500	300	6	1	6	36

$$f = 100 \quad fd = 103 \quad fd^2 = 505$$

Mean is $200 + 50 \times \dfrac{103}{100} = 200 + 51.5. = 251.5.$ Mean is \$251.5.

Standard Deviation is $50\sqrt{\dfrac{505}{100} - \left(\dfrac{103}{100}\right)^2} = 50\sqrt{5.05 - 1.03^2}$

$= 50\sqrt{5.05 - 1.0609} = 50\sqrt{3.9891} = 50 \times 1.9973$

$= 99.86$ dollars.   Variance $= 9972.75$ dollars^2

3.   Mean is 33.48 years, Standard Deviation is 10.83 years.

CHAPTER 10

10.3   1.   (a) 625   (b) 243   (c) 64   (d) 512   (e) 1

(f) 17   (g) 5.29   (h) 1

2.   (a) $25x$ (b) $9a$ (c) $11x - 4y$ (d) $14ab$

(e) $8a^2 - 2a$ (f) $10a^3$ (g) $12abc + 18bc$

3. (a) 56 (b) 80 (c) 60 (d) 124 (e) 320

4. (a) $30x^7$ (b) $24a^3b^3c$ (c) $4xy$ (d) $27y^4z^3/4$

(e) $60a^3b^4c^2$ (f) $4b^5/c^4$

5. (a) 39 (b) 15 (c) 2 (d) 58 (e) 1 (f) 76

6. (a) $8x^2 - 12xy$ (b) $30a^2 + 15ab$ (c) $9a^2 + 3ab$

(d) $2a^2 - 5ab - 3b^2$ (e) $8x^2 - 22x + 5$

(f) $6x^2 + 17x + 5$ (g) $9x^2 - 24x + 16$ (h) $16x^2 - y^2$

(i) $6a^2 - 5ab - 21b^2$ (j) $a^2 - ab + 4b^2$

10.7 1. 227

2. 0.88

3. -3.9168

10.9 1. Multiply by 7 then subtract 4.

Therefore add 4 then divide by 7; hence $x = \dfrac{y + 4}{7}$

2. $u = \dfrac{s - 16t^2}{t}$

3. $9x = 56$

4. $x^2 - 2 = 0$

CHAPTER 11

11.2 (i) 6 (ii) 5 (iii) 4 (iv) $^7/_{16}$ (v) $^{-3}/_5$

(vi) $^{17}/_7$ (vii) $^3/_2$ (viii) 13 (ix) 26 (x) $-4^1/_2$

(xi) 1 (xii) -97

11.6 1. $x = 4$, $y - 2$

2. $x = 5$, $y = 2$

3. $x = 2$, $y = -3$

4. $x = 2.5$, $y = 1$

11.10 1.   (a) -1 or -5 (b) 4 or -6 (c) -3 or $-\frac{1}{2}$

            (d) 1 or $\frac{5}{3}$ (e) $-\frac{1}{2}$ or $-\frac{2}{3}$ (f) $\frac{1}{4}$ or $-\frac{5}{3}$

            (g) $\frac{1}{3}$ or $-\frac{2}{5}$ (h) $\frac{4}{5}$ or $-\frac{5}{6}$

      2.   (a) 3 or -1 (d) 1.58 or 0.423

            (b) 0.275 or -7.27 (e) 1.72 or -0.117

            (c) 0.676 or -5.18

      3.   (a) no real solutions   (d) 2 different solutions

            (b) 2 equal solutions   (e) 2 equal solutions

            (c) 2 different solutions

CHAPTER 12

12.2 1.   (i) 5 (ii) 13 (iii) 10

      2.   (i) $\frac{7}{3}$ (ii) $-\frac{1}{9}$

      3.   (i) $6x^2 + 4x + 7$

            (ii) $6x^2 - 10x - 3$

            (iii) $14x + 10$

12.7 1.

x	-4	-3	-2	-1	0	1	2	3	4	5
$f(x)$	-35	-27	-19	-11	-3	5	13	21	29	37

      2.

x	-2	-1	0	1	2	3	4	5	6
x^2	4	1	0	1	4	9	16	25	36
$3x^2$	12	3	0	3	12	27	48	75	108
$-4x$	8	4	0	-4	-8	-12	-16	-20	-24
5	5	5	5	5	5	5	5	5	5
$f(x)$	25	12	5	4	9	20	37	60	89

3.

x	-3	-2	-1	0	1	2	3	4	5	6	7
x^2	9	4	1	0	1	4	9	16	25	36	49
x^3	-27	-8	-1	0	1	8	27	64	125	216	343
$-7x^2$	-63	-28	-7	0	-7	-28	-63	-112	-175	-252	-343
$3x$	-9	-6	-3	0	3	6	9	12	15	18	21
5	5	5	5	5	5	5	5	5	5	5	5
$f(x)$	-94	-37	-6	5	2	-9	-22	-31	-30	-13	26
$q(x)$	-5	-2	1	4	7	10	13	16	19	22	25

CHAPTER 13

13.4 1. 13.83

2. 66.31

3. 238.04

4. 52.4

5. 96.1848

13.10 1. Gradients are $-6, 0$ and 14.

2. 84 units^2.

3. 83.5 units^2.

4. $82^1/_3$ units^2.

CHAPTER 14

14.3 (i) True   (ii) True   (iii) True   (iv) True

(v) False   (vi) False   (vii) True   (viii) False

(ix) True   (x) True   (xi) True   (xii) False

14.7 1. (i)   { 3, 6, 9, 12, 15, 18 }

  (ii)   { 7, 14 }

  (iii)   { 3, 6, 7, 9, 12, 14, 15, 18 }

  (iv)   { } or ∅

  (v)   { 4, 5, 7, 8, 10, 11, 13, 14, 16, 17, 19 }

  (vi)   { 3, 4, 5, 6, 8, 9, 10, 11, 12, 13, 15, 16, 17, 18, 19}

  (vii)   { 4, 5, 8, 10, 11, 13, 16, 17, 19 }

  (viii) $\mathscr{E}$ or { 3, 4, 5, 6, 7, 8, 9, 10, 11, 12, 13, 14, 15, 16, 17, 18, 19 }

2. (i)   Set of squared numbers.

  (ii)   { 5, 10, 15, 20, 25 }

  (iii)   { 5, 7, 11, 13, 17, 19, 23 }

  (iv)   { 25 }

  (v)   { 5 }

  (vi)   { } or ○

  (vii)   { 5, 7, 10, 11, 13, 15, 17, 19, 20, 23, 25 }

  (viii) { 5, 6, 7, 8, 9, 10, 11, 12, 13, 14, 15, 16, 17, 18, 19, 20, 21, 22, 23, 24, 26 }

  (ix)   { 6, 8, 9, 12, 14, 16, 18, 21, 22, 24, 26 }

  (x)   { 6, 8, 10, 12, 14, 15, 18, 20, 21, 22, 24, 26 }

  (xi) $\mathscr{E}$ or { 5, 6, 7, 8, 9, 10, 11, 12, 13, 14, 15, 16, 17, 18, 19, 20, 21, 22, 23, 24, 25, 26 }

  (xii)   { 6, 7, 8, 9, 11, 12, 13, 14, 16, 17, 18, 19, 21, 22, 23, 24, 26 }

  (xiii) { 6, 8, 9, 10, 12, 14, 15, 16, 18, 20, 21, 22, 24, 25, 26 }

  (xiv)   { 5, 6, 7, 8, 10, 11, 12, 13, 14, 15, 17, 18, 19, 20, 21, 22, 23, 24, 26 }

  (xv)   { 6, 8, 9, 12, 14, 16, 18, 21, 22, 24, 26 }

(xvi) $\mathscr{E}$ or { 5, 6, 7, 8, 9, 10, 11, 12, 13, 14, 15, 16, 17, 18, 19,
                 20, 21, 22, 23, 24, 25, 26 }

(xvii)   { 6, 8, 12, 14, 18, 21, 22, 24, 26 }

14.9 1.   P ∩ Q = { 12 }

(P ∪ Q)' = { 1, 2, 5, 7, 10, 11, 13, 14 }

2.   (i)   6

(ii)   { Beryl, Desmond, Leo }

(iii) { Chandrakant, Elaine, Fearon, Henrietta }

(iv)   2

(v)   3

(vi)   { Adam, Beryl, Desmond, Guy, Idris, Jacquetta,
        Kathryn, Leo }

3.   (i)   { 14, 18, 21, 24, 28 }

(ii)   { } or ○

(iii) { } or ○

(iv) $\mathscr{E}$ or { 13, 14, 15, 16, 17, 18, 19, 20, 21, 22, 23, 24, 25,
        26, 27, 28 }

(v)   { 13, 15, 17, 19, 22, 23, 25, 26, 27 }

(vi)   { 28 }

(vii) { 24 }

(viii) { 13, 15, 17, 18, 19, 22, 23, 25, 26, 27 }

CHAPTER 15

15.5   1.

A	B	A.B	A.B	(A+B).(A.B)
0	0	0	0	0
0	1	0	1	0
1	0	0	1	0
1	1	1	1	1

2.

A	B	A.B	$\overline{\text{A.B}}$	B+$\overline{\text{A.B}}$
0	0	0	1	1
0	1	0	1	1
1	0	0	1	1
1	1	1	0	1

3. C is $\overline{\text{A}}$

D is A+B

E is $\overline{\text{A+B}}$

F is $\overline{\text{A}}$. $\overline{(\text{A+B})}$

A	B	$\overline{\text{A}}$	A+B	$\overline{\text{A+B}}$	F+$\overline{\text{A}}$.$\overline{(\text{A+B})}$
0	0	1	0	1	1
0	1	1	1	0	0
1	0	0	1	0	0
1	1	0	1	0	0

4. D is A.B

E is B+C

F is $\overline{A.B}$

G is $\overline{B+C}$

H is $(\overline{A.B}) . (\overline{B+C})$

A	B	C	A.B	B+C	$\overline{A.B}$	$\overline{B+C}$	H
0	0	0	0	0	1	1	1
0	0	1	0	1	1	0	0
0	1	0	0	1	1	0	0
0	1	1	0	1	1	0	0
1	0	0	0	0	1	1	1
1	0	1	0	1	1	0	0
1	1	0	1	1	0	0	0
1	1	1	1	1	0	0	0

5. (i) $\overline{A.B.C.}$

(ii) $\overline{(A.B) + (B.C)}$

(iii) $(\overline{A+B}) . (B+C)$

(iv) $\overline{(A+B) . (B+C)} . (A+C)$

6. (i) $A+\overline{B}$

(ii)   B

(iii) B

(iv) $B.\overline{C}+\overline{A}.\overline{B}$

(v) $A+\overline{B}.C$

CHAPTER 16

16.6   1.   If the manager is A and the assistants are B and C then the requirement is met by A.(B+C).

2.   A.B.C.

3.   A.B+A.C+B.C

4. $\overline{D}+D.\overline{B}.\overline{C}$.

CHAPTER 17

17.4   1.   (i) 6.8   (ii) 5.3   (iii) not defined

2.   (i) Jan   (ii) Harriet   (iii) Liang   (iv) Philip

(v) $T_{4,2}$ (vi) $T_{4,5}$

17.10   (a) $\begin{bmatrix} 6 & 7 & 8 \\ 9 & 4 & 13 \end{bmatrix}$ (b) $\begin{bmatrix} 6 & 21 & 9 \\ 24 & 3 & 18 \end{bmatrix}$

(c) $\begin{bmatrix} 0 & 14 & 1 \\ 15 & -1 & 5 \end{bmatrix}$ (d) $\begin{bmatrix} 6 & -7 & 7 \\ -6 & 5 & 8 \end{bmatrix}$

(e) $\begin{bmatrix} 50 & 45 \\ 98 & 21 \end{bmatrix}$ (f) $\begin{bmatrix} 4 & 41 & 12 \\ 44 & 19 & 36 \\ 48 & 60 & 48 \end{bmatrix}$

CHAPTER 18

18.5   **2.**

−1	Accounting	2
3	Packages	6
8	Include	4
−1	Not	5

−1	Only	−1
7	The	10
−1	Software	−1
−1	Developed	9
−1	For	−1
−1	Users	−1

3.

−1	2	Accounting	−1	8
3	6	Packages	1	7
8	4	Include	2	4
−1	5	Not	3	5
−1	−1	Only	4	2
7	10	The	2	10
−1	−1	Software	6	6
−1	9	Developed	3	9

| −1 | −1 | For | 8 | 3 |
| −1 | −1 | Users | 6 | −1 |